FINDING THE FOREST

Stacey,
More reminders of
the good life in
the Lake States.
Enjoy

Peter

Finding the Forest

The Initiation

Peter P. Bundy

with Illustrations by Mary Sandberg

Masconomo
Forestry

In loving memory of
Harvey H. Bundy, Jr.

Table of Contents

Introduction

WELCOME TO A JOURNEY into the future, a journey which originated in the past. Over the years I have recorded notes and observations in a series of forestry journals I keep. Sometimes my journal notes have been practical in nature, based on concrete situations and specific events. Sometimes the observations have been more aptly entitled musings as I followed the national debate on conflicting methods for managing our nation's forests.

At first I did not expect that these notes would germinate into a book. They were private thoughts for my own education and memory. Then I began sharing some of the entries with friends. They encouraged me to "go public" with some of these essays. I reworked a couple of them and when the first two submissions were accepted for publication, I was flattered. Perhaps my friends were on to something?

Now the real work began. I organized the entries with Norman Maclean's *Young Men and Fire* in mind. Maclean has a gift for writing non-fiction with the arc of a dramatic story. I cut out much of the personal stuff and found a rhythm with a beginning and an ending. It is an offering.

As we approach the twenty-first century we are at a critical junction in our attitudes toward uses of the forests we are blessed with. The old models and methods of using or "exploiting" the forest are under fire. The new methods of "ecosystem-based management" are untested and largely theoretical at this time. Some wish to return to the past and simpler times. Others wish to throw out all of our experiences of the past and start with a clean slate. What are we to do?

In these pages we will follow one forester's journey. He was not born in the woods, but arrived as an outsider, an urban refugee. Over the years, as he built a home and trained himself, his experiences began to form a new image. The image grows like a photographic plate as it is exposed to time.

This story is a series of vignettes which inhabit their own space. Taken alone, the vignettes are of interest. Woven together, they offer a different vision of the forest. It is a vision which challenges many of the assumptions we have about trees and forestry practices, assumptions about "preservation" and "utilization." I believe it is time for us to carry the debate about our forests and their future to a new level. To do that, we first need to improve our vision. These pages initiate that journey.

Building a Beginning

Gravel Roads

MOST YANKEES COME FROM THE NORTH. Their heritage is in the north. Their hearts are in the north. Their spirit is in the north. The north is where their dreams begin.

In the spring, a solitary traveler heads north on a blacktop, four-lane highway, an insulated ribbon of asphalt. The stereo is on. The windows are closed. The landscape is out there—a hundred yards from the glass—like the view from an express train. It is all very safe looking, or so it appears from the armchair highway.

Slowly the landscape changes. The farms are fewer. The silos and barns are older. Hay bales girdle the homesteads. The towns become smaller and the traffic thins. The forest appears. It all happens in slow motion.

The four-lane expressway ends. With two lanes, the traffic from the other direction becomes real. I notice the semi-trailer fully loaded with lumber bearing down from the north. The houses are closer too. There are cars in the driveways. An old pickup pulls out in front of me. Old pickups never go over fifty miles per hour. I hit the brakes. The landscape closes in.

Thoughts of the past drift up from cracks in the asphalt. Memories of successes and failures rise and fall. Faces float down from the sky, carrying burdens and blessings. There is exhilaration and fear in the air. A journey into the unknown is about to commence. Only the whine of rubber on concrete greets the ears, and a highway that narrows, and narrows and narrows.

Finally, the road turns gravel. All real roads are gravel. When I hit gravel, I know it. The car tells me. The ride changes; the small pebbles kick up. The dust billows forth on a dry day. Anything can happen.

ℬ   •   ℭ

Childhood Shelter and a Chorus

LONG AGO, WHEN I was a child, the solemn line of evergreens outside my Massachusetts window was a refuge. They were never known as anything other than pine trees: not white pine or jack pine or Scotch pine. Only later did I discover they weren't pines at all but Norway spruce. Spruce offer shelter.

Intertwined with the branches curved and wide, their shade was a place to crawl into, a secret world of soft needles, pitch, and solitude. It was away from skirmishes with my brother, away from parental expectations, away from poor school days, away from all childhood disappointments.

To these trees I would come to seek solace after some typical childish episode. Here I found calm, security, peace: the smell of the pine needles, the sound of the wind whistling through, the touch of the soft ground, the darkness of shade on a bright fall day. It was a natural place to go to be alone, just as the highway north is twenty-five years later.

My car chatters to a halt. It is spring. Late April to be specific. The air blows warm from the south. Love songs of migrant birds are everywhere. The ground, however, is still frozen, frozen and almost lifeless as the boots pass across last year's debris. I am greeted with dead leaves, broken branches felled by winter winds, old raspberry brambles twisted into so many knots.

But I am not looking down. I walk toward the lake and the spirit of the songbirds. The hepatica peeps through the leaves, full of promise. The swollen buds of the aspen appear on the side of the trail. A strange sound beckons from afar. It is a new sound, and it is still distant, like a chorus of children across the water.

Slowly the song grows as the trail winds toward the pond and the lake beyond. Clearly the voices have life. But what kind of life is it that grows out of still-frozen ground?

A pond appears through the willows, and the chorus becomes stronger. Piercing and rhythmic, a cacophony of voices assaults the eardrums and pulses into the brain. The mind is baffled. What earthly creatures call out from the depths of the swamp?

A northerner will know the answer to this question immediately. But for one who has not seen the power of spring in the northland, it is like stepping into the Paleozoic. This choir does not come from humanity or the present urban landscape. An old Greek chorus, with millions of years behind it, it arises from cold waters of the past.

It is the frogs. The frogs, climbing out of the mud of a long dormant winter, announce, proclaim, shriek their strange harmony. Spring is here. Expectancy is everywhere.

Esden Lake

TODAY, ESDEN LAKE is a small shallow lake in Central Minnesota. Carved during the last great ice age, 8,000 to 14,000 years ago, Esden Lake is not a spectacular body of water, not a Lake Superior or a Tasman Sea. Instead, it is intimate. Three to four hundred yards in width and almost three-quarters of a mile long, it is modest in its demeanor and stature. And yet, in its modesty lies its charm.

The substantial expanse of blue reflects the Eastern sky in all its many moods of light. Calm gray, shimmering cobalt, mottled white; these are reflected from this bowl set gently into the north woods landscape.

There are moments when the surface of the lake is a perfect reflector of the morning sunrise and other moments when whitecaps froth the shoreline. A thousand winds play on the surface and dance about. The birds know these winds and enjoy them. The loons call dramatically from center stage. The buffleheads float serenely in the shadows of the windward calm. The pileated woodpeckers scoop V's across the surface, and high overhead a bald eagle twirls and soars to the rhythm of the dance.

I watch from the shoreline. Sometimes the dance is crowded with activity. More often the lake is a solitary place. Few boats ply its waters. A canoe appears briefly on a summer weekend and is gone. Silence returns.

Most of the year, Esden is a peaceful lake. The western shore is defined by a thirty-foot ridge. This "esker" of glacial gravel and rock, which winds up from the south, was formed by a stream flowing beneath the glacier as it melted thousands of years ago. It and the lake are the defining topographic features of the local landscape. It is a perfect spot to stop and stay. Someday I will call it my home. But before I can build a home, I need access and a road.

A Roadway and a Fedora

USUALLY THE WOODS are quiet and cheerful early on a spring morning. This morning, however, there is a menace and hum in the air. You can feel it from a quarter mile away—the sound of a large engine in the woods—an alien sound.

The first time you experience this yellow creature at work up front and close, you will remember it, especially if you're standing in its path. The ground underneath seems to move and tremble. Those sturdy ironwoods and bur oaks bend like feathers in its jaws. The ground is turned up, revealing dark damp earth in ochre, and umber. Boulders too large for a man to lift pop up like gum balls. Stand back! The machine moves on.

When I reach the brushed-out trail, I stop. This is no ordinary man on an ordinary machine. The yellow metal monster lumbers through the woods in front of me, its power obvious and striking. It fears nothing. The forest parts before its steel blade like so many popsicle sticks stuck in a sand box. The sound of destruction overwhelms the air: tree roots breaking, branches tearing apart, debris plummeting to the ground. It is an act of man with a steady heartbeat.

But there is a method to the madness. Trees are lined up gently like fallen soldiers. Gullies and hummocks disappear. Ditches appear. A road grows out of the ground. A level surface can be imagined. The man and his machine move on to the next 200 yards of the path.

Who is this whiskered man at the controls? He passes me with a smile and a wave of the hand. The machine slows its pace and pauses. The growl shifts to a purr. He steps down—the image of a cowboy from the past.

His hat is the first thing I notice. It is a wide-brimmed, brown-leather fedora, and it appears to have been through a world war. Torn, stained, turned down in front, it has grown whiskers like the man who wears it. Probably he sleeps in it, I think to myself.

The man is smaller than I expect. With a light frame, he's a wiry spark plug with a toothless grin a mile wide. "Well," he says, as he surveys the landscape he has wrought. I am speechless. My eyes must show it. I try to say something intelligent, but out comes, "You got an early start." He only nods and looks back at his machine. I try again, not trying to sound smart. "Pretty impressive!" He looks at me and smiles again. Will he speak?

"She's got a lot of hours, but she gets the job done." He refers to this yellow monster in the woods as his baby. His voice is soft and gentle. His eyes sparkle with delight at his work. I know nothing about machines, let alone bulldozers. It's my turn to nod.

By late that same afternoon, a new road takes shape in the forest. The dark fresh earth is smoothed and ditched. Sentinel trees are piled by the road bank, awaiting a chain saw and their next life as firewood. The madness and chaos have been transformed into a newly tilled carpet of earth, compacted and winding its way into the wilderness.

There will be another day for my friend in the fedora to finish the ditches, to place culverts over low areas, to smooth out the crown. But the new road is passable. I cannot believe the change. My small blue car sits poised on the township road beyond. It is time for a drive. Do I dare?

This is the first pass of wheels over this new ground. The earth is still soft and breathing of the forest. Will it hold the weight of rubber and steel? I inch a few tentative feet and step out to examine the depth of the ruts I have left. I am uncertain if I can trust this new highway. Perhaps it will need time to settle. Surprisingly, the tracks are light, at most an inch deep.

It is time for the virgin run. The door closes and the gearshift goes into first. I am moving. Surprisingly, first gear seems too fast. The new landscape slips by before I can digest it. Around a bend I slide. The nose tips toward the ditch. This is no time to view the scenery. I concentrate on staying afloat.

The remainder of the first run is a blur. The wheel in my hand seems to guide itself over hummocks and around bends. Finally, I cross the final stretch to the ridge, and there in front of me is the lake. I have made it into my new homesite. I pull to a stop, and my new friend with the fedora smiles.

Building a Home

BUILDING A HOME, even a small one, is a ritual. You become part of the landscape. Like a tree, you place roots into the earth, especially if you dig the foundation yourself. As your shovel hits rocks and roots, you begin to realize what the landscape is made of.

Like most novices, I had no concept of what I would face when choosing to "build myself." Had I known the reality of the physical pain, the setbacks and the heartaches, I probably would have turned the car south, back to the safety of the city. But that is hindsight.

℘ • ℘

I had no prior experience in building, no experience with rebar, mortar, or generators, with plates, joists, or rafters. Embarking on a home-building project is an act of faith and foolishness. Faith, because I put my trust in forces larger than myself. Foolishness, because I might not survive the process. Many home-building projects have ended in injury or abandonment. If the building gods are on my side, mine will succeed.

There are a few ways to start off on the right foot. One is to have legal access to the house site and to check local building codes. Another is to spend time on the design and to read books by others who have gone before. And, of course, you need to build within the budget. The list is long and could be longer. Let's see how I did.

a) Access: I did not have legal access. I had a temporary access across a neighbor's land. Strike one.

b) Building codes: Local ordinances would not let me build on the ridge due to zoning regulations and a two-hundred-foot setback requirement from the lake. I had to initiate a variance procedure, which is a slow, risky process. Strike two.

c) Design and books: I made many drawings and even a cardboard mock-up of the home design. I also had a voracious appetite for all kinds of manuals and books on the subject. Ball one.

d) Advice: I hired a couple of friends who were professionals to help solve problems and to work with me for two weeks to frame it up. Ball two.

e) Money: I was designing a roughed-in structure for $20,000. The money was in the bank. Ball three.

Although I was off to a rocky start at the plate, I pulled up to a full count as the summer weather appeared on the mound.

೫ • ೞ

There is no need to bore the reader with all the ten thousand details of designing and constructing a house. In my case, due to the delay in zoning variances, I built a small eleven-by-sixteen-foot cabin first to get out of the rain and mosquitoes. The first night I had a roof over my head, a thunderstorm came through and blew down my old tent. I was grateful for the sturdier shelter.

As the summer wore on, the local planning commission gave me permission to build on the ridge. My friends arrived with radial arm saws and power augers. The footings went in, the concrete became lighter. Slowly a structure went up on the ridge overlooking Esden Lake. It was exhausting and exhilarating at the same time. At night I didn't have the energy to think or to worry. I would collapse into a deep sleep and dream of two-by-six stud walls.

Building a home is also a ritual of seasons. As the footings, foundation, and frame grow, the clouds change. The sun moves in the sky. At first it is higher with the heat and the long evening light. Then the days start to dim, and the cool winds arrive from the north. The race to close up and prepare for winter begins in earnest. The roof goes on with relief. The windows go in with delight. The insulation is inserted with itchy unpleasantness. Finally the wood stove arrives.

And in all this, the woodpile is still waiting to happen. There will be time to rest by the hearth, but that time is in the future. The squirrels and chipmunks on the nearby branches and logs chatter and prepare their nests for the long nights and cold days. I do the same and prepare myself to become one of the inhabitants of the woods.

Water and Witching

IN THE CITY, water comes out of a tap. To take a drink or a shower, I turn on the faucet. It is simple and magical. When I decided to build a house in the woods, I needed water, but it was not as simple as turning on a tap. The first step was to find it.

Terry is the neighborhood "water-witcher." I invite him out to Esden for a stroll around the house site. He is an unassuming guy with a hearty laugh. At first glance he does not strike me as unusual in any way. This is an illusion.

Terry unfolds his pocketknife and wanders off into the woods. A few moments later he reappears with a couple of small willow branches between his fingers. Silently he goes to work on these twigs and whittles them into forked tongs about two feet long, with a Y crotch and carefully whittled prongs.

He rises from his knife work and balances the prongs on his upright fingers, one in each hand, held outstretched in front of him as if in prayer. He nods his head in satisfaction. We are ready to go exploring.

On this ground, finding water is not the problem; there is water everywhere. But finding the veins at a reasonable depth for hand digging a sand-point well—that is the challenge. Terry starts walking with his wishbone willow held upright near his chest. He tends to walk in arching circles, balancing the willow aloft like a top.

Every few minutes during this wandering, the stick mysteriously starts to turn over. Down it plunges toward the ground. Terry stops at the spot. "Here," is all that he says. I place a wooden stake on the spot, and we move on in our circling pattern.

We wander the site for ten or fifteen minutes in this fashion. The willow leads the way. I follow with the stakes, still a bit skeptical of the man and his methods. I watch for clues that the willow will drop, but there are none I can discern. No change of position, no moving of the fingers. I even ask to hold the willow in my fingers, but it falls over before I even take one step. Terry smiles.

With four or five stakes in the ground, Terry isn't finished. "Which spot do you want?" He gives me a choice in this matter. I pick out the stake nearest the future kitchen. Now the real magic begins.

He kneels by the spot and assumes a still posture with the willow balanced over the ground, as though he is preparing to roast a marshmallow on an imaginary fire. There is a pause. Motionless, he waits. Then, the willow starts to bob. It wags up and down, a mysterious force pulling it down and letting it go. Terry is counting. Finally the bobbing ceases. Terry stands up. "Thirty-seven," he says. I am puzzled.

Later, by the side of his pickup as I hand him his payment of a twelve-pack of soda, I have the courage to ask what the thirty-seven means. "You should hit the seam at thirty-seven feet." He smiles again and glances at his fishing boat hitched to his truck. He is itching to be off on the water, a water-witcher indeed. I let him go without further questioning.

Two weeks later, the well-driller hits water at thirty-eight feet.

The Flight of the Snowflake

THERE COMES A TIME when I realize that I am alone in this new landscape. I have left my childhood behind. I have set off on a path of my own choosing with only my heart and mind to guide me. The trail has not been blazed. What happens during this journey will shape the remainder of my life. I may succeed. Or fail. Most likely I will do both as I reach into the earth for nourishment.

For me, this time arrives early one October morning as I awaken from a deep sleep to find two inches of white on my doorstep. The first snow has silently arrived in the night.

The house is half insulated. I still have no electricity. My wood stove is cold. My woodpile is non-existent. I sit up and stare in disbelief at the whitened world around me. Is this a dream? I shiver.

Somewhere in the corner, a clock ticks, reminding me that I am still alive. For all of my labors of the past six months, I am at ground zero for winter.

I rise and find some kindling two-by-fours. I split them, crunch up old newspaper, and start a small fire. Just the light of the flames dancing in the cast-iron frame raises my spirits. It is too late to quit. It is time to hibernate, time to stop fighting the darkness. I insulate the walls and close the door. And when the house is silent, when the last rays of light disappear from the western November sky, I sit in my favorite corner and draw strength from the pines outside my window.

There they stand, tall, silent, serene soldiers of green. They know about darkness and contemplation. They have stood in the same spot for nearly a hundred years, and they are healthy and strong. I am becoming fearless, like the forest. I breathe deeply and hear my heartbeat against the walls of oblivion. It beats on. And now, as I glance at what my hands have wrought, for the first time I can rest.

୬ • ଓ

I stare out the window of my new home. I eye the gentle snowflakes as they appear out of the gray mist and then disappear again into a snow bank. Their life span as individuals is brief. A few seconds from cloud to ground. For this short time these frozen crystals have their own path, their own shape, their own beauty. In a sense, they are free from their brethren and from the primordial soup. Free to choose their own journey. This freedom, this individuality is fleeting for the snowflakes outside my window. Soon they are joined again to the history of their kind. Soon they will be part of the collective past.

My journey as a human is similar, though the span is a bit more extended. I am conceived and raised in the collective culture of my family and my country. Most of my days are spent with my family of the past or my family of the future. For a short span, however, I am on my own. I am alone as I drift. I have my own identity, my own physical body, my own flight. And to a certain extent, during this brief window I have the ability to make choices that will shape my future. The choices I make will seem like simple ones. Who are my friends? How will I spend my Sunday afternoons? Where shall I live? But these are the simple choices that will shape my future. The friends I meet may become my family. My hobby may become my career. My mailing address may become my community.

After my home has become a shelter, after the wood stove is purring out heat, after the lights are on and the well is pumping water to my kitchen, I have a moment of choice. A moment to decide how to spend the quiet evening. I glance outside at the snowflakes again and follow their brief flight. On and on they pass into the night. My journey into the forest is just beginning.

Entering the Forest

Itasca and Cloquet

THERE ARE MANY PATHS into the forest. One path winds south from my door along the ridge. This old deer trail is a path for observations. Here I have watched the beavers at work, followed fox tracks in the snow and enjoyed the trillium in full bloom. This trail winds in a slow loop and brings me home again with reassuring ease.

There are other trails into the forest that lead in more challenging directions. One trail I chose to follow led me out of my woods and into a wider world. To earn a living in the forest, I signed up to return to school. The University of Minnesota offers two forestry field programs, Itasca and Cloquet.

Itasca has a mythological tone. It is the "verITAS CAput" (Latin for "true head") of the Mississippi River, a pristine lakeshore surrounded by two-hundred-year-old red and white pine. It is one of the only areas of northern Minnesota to have escaped the Swede-saw of the late nineteenth century.

It turns out that early explorers had a great deal of difficulty finding this place. The Mississippi River does some surprising things in its upper watershed. It turns east and runs into many lakes. It turns west and runs into more lakes. Finally it turns south. The explorers kept traveling north to find its source, an old obsession among western explorers. Never mind that the Missouri River tributary might be more aptly called the headwaters. Never mind that the Minnesota River is significantly larger and longer than the Mississippi at their junction in the Twin Cities. These explorers wanted a river that came from the north. Head north to the source—north, it turns out, to a small lake tucked away in northern Minnesota: Lake Itasca.

Here, where the first state park in Minnesota was created amidst a small fragment of original white and red pine forest, I was sent to a month-long initiation rite. The Itasca session. What a setting for a school.

The drive to Itasca takes me away from the hustle and bustle of the city, away from the interstate, away from shortening nerve endings. The cool northern forest is verdant and lush in late summer. As I approach, I recognize this as a different place. The pine trees are older and taller, and the forest has a different feel to it. Later I will learn why. For now it is enough to settle in a cabin by the clear blue lake and wander along a new trail on the quiet shoreline. It is a promising start.

For more than a month I live, eat, sleep and study in this romantic setting. Rising early to study botany in the field, the names of herbs and forbs fill the brain: *Amphicarpa, Ariseama, Atrorubens, Auralia nudicaulis.* . . . This is a language from another time. I don't read a newspaper or listen to the radio. The sounds of wind, wolves, and owls suffice.

In the afternoons, I burrow into the soils, digging to see what this planet looks like beneath the surface. It is a surprising journey, filled with marvelous colors and strange names like *Spodosol, Alfasol, Bhir*. It all means something, and I try to remember what I see. I will be tested on it.

And then there are the trees: three-hundred-fifty-year-old white pines, three-hundred-year-old bur oaks, two-hundred-fifty-year-old red pines. These trees have all been alive longer than the United States of America. They have something to offer.

The days stretch into weeks in this intense environment. I'm too tired in the evenings to appreciate the serenity, with the botany quiz in the morning and the ecology paper to write. Later, from a distance, I will look back at the people I met on these paths, at the evening light and at the pines. I will remember my first ecology lab, my introduction to forest succession and my attempt at habitat classification. I will leave Itasca with a different set of eyes and my first experience in how the northern forest functions.

๛ • ๙

Cloquet is the other side of the coin. Hidden on the sand plains west of Lake Superior, Cloquet is a forestry research station for the university. Here scientists study growth rates of pines, establish seed orchards with improved genetic traits, experiment with differing methods for releasing preferred conifers on cut-over lands.

At Cloquet the concentration is on traditional forestry practices. I learn how to measure tree volumes in cords and board feet, how to run a traverse with a compass, how to put in sample plots for a forest inventory. I learn what mean annual increment, the central limit theorem, and aerial photo displacement mean. I run regression analysis for paired plot research designs and set up a silvex exam by United States Forest Service standards.

It is a rigorous three months, with simple bunk beds and three square meals a day. There are a couple of hours of classroom work in the mornings, with the remainder of the day in the field. I might find myself running compass lines in the snow, with cold wet boots and less than adequate cover gear. Or I might find myself touring a modern paper-making plant with aspen logs going in at one end and two-ton rolls of glossy printing paper coming out the other. I might spend the morning drawing maps and pouring over air photos and the afternoon trekking through the lowland conifer swamp that looked like an easy traverse on the photos. Most likely, however, I will find myself in the quiet forest with my clipboard in hand measuring basal areas with my prism, taking increment borings of tree cores, utilizing my clinometer to estimate tree heights. I begin to see the variety and the texture in the forested landscape in a new way, just as I did digging in the earth at Itasca.

During the three months at Cloquet, I build new friendships with the small group being tested. The cabins where we sleep are small and rustic. The meals we share are communal and hearty. There is only one telephone line out of the compound. The outside world recedes into anecdotal history, as it did at Itasca. The present becomes the weather forecast, a dry pair of boots, and a full lunch pail.

      ഌ   •   ര

If Itasca is the John Muir session, Cloquet is the Gifford Pinchot session. Muir and Pinchot were both trained as foresters. In fact, in their early years they were friends. Then their visions took them down different paths, and the result is still very much in evidence today.

Muir was a westerner. When he went into forestry, he naturally enough chose the U.S. Forest Service as a place to start, but he became disillusioned with public sector forestry. He was disturbed at what was happening to the redwood forests of northern California, and he was disturbed at what he saw after the loggers passed. He saw silted-up streams with the salmon gone. He saw hillsides turned into mud slides. Most of all, he saw the voracious appetite for wood that the new American culture had developed.

His response was to fight back to retain wilderness. Fight for the California condor and the golden eagle. Fight for the back country, for Yosemite, for the Sequoias. Muir, in this sense, can be considered the father of the preservationist philosophy.

Pinchot, on the other hand, was an easterner. He was well connected, and he preferred to spend his time in the halls of Congress.

Pinchot also had a vision. He had a vision of forestry as practiced by the Europeans. He too was shocked at the waste and destruction of the "slash and burn" logging prevalent at the time. But his response was different. He wanted government to step in and manage the great western lands responsibly. He thought that government had a right and a responsibility to protect what the private sector was ignoring: the future wood and water supplies for a growing country. In his opinion, the private sector could not do this with its appetite for short-term economic benefits.

Pinchot was a people person, a talker and a politician. He lobbied and he lobbied hard. He also knew the right people, Teddy Roosevelt for one. Roosevelt was an outdoorsman who appreciated Pinchot's vision. Together they put together the largest land acquisition program in our nation's history—the backbone of today's 190 million acres of national forest. When you walk in a national forest or camp in a national forest campground, you can thank Gifford Pinchot.

Pinchot and Muir had different visions, but both of them have given us a great deal. While many of the conflicts today result from clashes in these philosophies, they share more than is appreciated. Like the two great political parties in America, without both, there would not be a democracy.

It has been almost one hundred years since Gifford Pinchot and John Muir laid the groundwork of modern forestry and wilderness management.

Since the 1980s, advocates of the two philosophies espoused by Pinchot and Muir have unfortunately become entrenched in bitter warfare over how to manage our national forests.

When Ronald Reagan and Interior Secretary James Watt were at the helm, national forests were managed primarily for timber production. Economic growth was the model and the United States Forest Service responded by "getting out the cut." More than ten billion board feet per year were harvested from national forests to build houses and produce paper for American households.

In the 1990s, the pendulum swung. With Bill Clinton and Interior Secretary Bruce Babbitt in charge, the preservationist approach, which had been gathering momentum, received full favor. Harvest levels on national forests were reduced sixty to eighty percent. Endangered species like the spotted owl were given priority for management. Water quality, recreation, and biodiversity issues shared the spotlight.

What does the future hold? Will this seesaw pattern continue? Will the forces behind exploitation and preservation continue to divide the popular airwaves? Will America's public forests continue to be the battle-ground for ideological and value-driven differences? Or is there another path?

୨୦   •   ଓ

What do we really want from our forests? Do we want raw fiber to build our houses and to print our newspapers? Do we want clean water? Do we want to be able to enjoy a grove of five-hundred-year-old Douglas fir? Do we want to fish for salmon? Do we want to have our cake and eat it too? If our goal is a sustainable forest resource with maximal long-term benefits for society, there is a need to find a balance between traditional economic and modern environ-mental benefits.

On the one hand, the forest is more than the bucolic place about which Thoreau, in his limited observations, wrote. After all, he was at Walden Pond less than a year before retreating to urban Cam-bridge. To the extent that the preservationists view the natural world as an idealistic vacationland, they express ignorance about the true functioning of the forest community.

At the same time, we must also understand that nature's bounty is limited. We do not need to repeat the "cut and get out" mistakes of the past. To the extent that the forest-products industry views our public lands as their wood box, they ignore other important benefits which the public forests provide our nation.

I believe there is another path that draws from both sides of this history. It is a path of understanding that we and the forest are not separate but equal. If we thrive, we can help the forest thrive. If the forest thrives, it will help us prosper. To follow this path, we first need to learn the language of forestry.

Silviculture

AT FORESTRY SCHOOL there was a course entitled silviculture. It sounded intriguing, so I signed up. Our professor started us out with a simple enough definition: Silviculture is the art and science of tending forest trees. Or, more simply put, it is applied ecology. That was enough to get me interested, and ten weeks later I was hooked. I started talking to my friends of shelterwood, seed tree selections, systematic thinning; of habitat classification, indicator species, and succession; of *Picea glauca, Pinus strobus, Quercus rubra.* My friends looked at me as if I had gone off the deep end.

In a sense, all of us are silviculturists at heart. Every time we walk in the woods and notice the ferns shading the ground, the mushrooms growing on the boles of the aspen, the balsam fir seedlings at our feet, we are participating in the world of silviculture.

Why are these seedlings growing here? How much light does this sapling need? What makes this bole straight and this one forked?

The art and science of silviculture is, first of all, an understanding of the complex natural processes by which forests grow, mature and decay. Some call this forest ecology. Certainly foresters, before we can prescribe plans or treatments for some desired future condition, need to observe how nature works. Silviculture, however, does not stop at observational ecology. It takes an active hand in shaping the forest of the future. We participate in the process and influence the course of events. We experiment.

Some of the things we try don't work very well. Then we try other things until we find some course of action that provides greater benefits to more people and to the health of the forest as a whole. This is the basis of silviculture. It is both an art and a science.

In traditional forestry circles, I hear a good deal about the scientific part of this equation. What is the current annual increment (the growth rate) of red pine? How much calcium does the root system of aspen uptake, and where is it stored? What is the ratio of red oak advance regeneration height growth to seedling height growth? These are important questions, and the pursuit of answers is worth the time-consuming measurements and calculations. Often this occurs in the more rarefied arena of research, where regression equations and statistical significance are the measures of success. Many scientific studies, however, are limited by the questions that are asked. If we don't ask the right questions, the answers may be meaningless.

Here is where the artist is sorely needed in forestry. The artist sees the landscape in a different light and asks different, more difficult questions. The artist has the vision to see possibilities for the future forest. It is the artist who is able to synthesize past information in new ways, to experiment with different techniques of growing and tending trees and to offer new solutions to old problems.

In tending our forests, we need the artistic vision at least as much as the scientific one, perhaps more. We need it not because it is more important than the scientific view, but because in forestry it largely has been neglected. We need to encourage foresters to try new solutions, to ask more questions—in short, to be more creative.

The artist notices shapes and patterns in the landscape and tries to make sense of them. The artist merely observes and is able to pause and watch and synthesize.

What happened to the acorns that fell from these trees last fall? Why are these red oak seedlings so numerous in this small cove when there are none 100 yards to the north? Where did the blister rust disease enter this tree and why didn't it affect its neighboring pine?

The art and science of silviculture offers a method to integrate the artistic and scientific worlds, to integrate the subjective with the objective. Through integration we reach for new solutions for the future forests of our planet.

The Ecology of Oak

WE ARE BLESSED WITH OAKS in the north woods. They live along the edge of the well-traveled highway, their branches sheltering us when we travel. In the country, they watch over pastures where the cattle graze. In towns, they hover over the playground at the elementary school. They stretch across hundreds of thousands of acres of second-growth forest that make up the Lake States landscape. Oak: the word conjures up images of strength, integrity, and beauty.

Today, as I walk in the woods and arrive at an area covered with oaks, I pause. How did these trees come to grow on this knoll? I lean against one of them and look for clues. Nearby are old pine stumps. The stumps are large, twenty-four inches or more. They are hollow, and they are charred. Fire scars and charcoal mottle them.

When the white pine groves were harvested about a hundred years ago in the Lake States, the lumber industry did not pay particular attention to the oaks in the understory. It was the overstory they were excited about: tens of thousands of board feet of high quality pine on each acre. But there, in the half-light beneath the canopy, were thousands of small oak seedlings and saplings. The pines were felled and the canopy opened up. Light reached the floor full blast. The oak seedlings responded, and so did every other sun-loving species in the neighborhood. Raspberries, aspen, birch, hazel . . . the list is long. The mad rush for the light began.

Unfortunately, not all of the pines were gone. In their haste, the loggers left behind everything that would not make a board. All the tops and limbs and branches were there, drying out and waiting for a spark. A tinderbox was created by thousands of acres of pine waste products. Fire hazard was not a known concern. There were more houses to build in the growing Midwestern towns and more trees to cut further west.

Then, the fires struck, as certainly as the dry years which preceded them. The skies turned dark, lightning flashed and flames lit the air. Nature, in her fury, tore through these slash-pine landscapes, leaving smoke and charcoal in her wake. The landscape turned black. The depression of the 1930s hit. The blackened earth was left for lost.

But mother nature was not finished. After the holocaust, after the destruction and death, quietly, like the sunrise, she went back to work. The maple was gone, its thin bark no match for the heat. The hazel and raspberries and ironwood suffered the same fate. The oak tops were charred and dropped off. But beneath the soil, the oak roots were alive: alive and ready to send up new sprouts. These seedling sprouts rose quickly with their large intact root systems. They encountered no ironwood or aspen or maple to overtop them. These sprouts were thick and had to outpace each other, like the ten thousand runners at the marathon. A new oak forest began its journey into the canopy.

℮ • Ⅎ

Foresters and ecologists refer to this process as secondary succession. Primary succession occurs on bare mineral soil, on agricultural fields, on post-glacial debris, on old mining sites. In primary succession, pioneer plant species such as aspen and birch colonize an area where no forest preceded them.

Secondary succession occurs when one type of forest is replaced by another. This usually occurs over a long period of time. Secondary succession is less predictable than primary. Many subtle influences and changes alter conditions on the forest floor and affect which species survive.

In the case of the oak forest where I stand, fire is the key disturbance that influenced secondary succession. The oaks are now entering their prime years of growth, some seventy years after a forest fire eliminated most of their competition. Nearby, however, secondary succession is different. Another severe disturbance, a windstorm, blew through forty years later and knocked down most of the oak. There, maple seedlings and saplings, were released into the sudden light. Secondary succession is proceeding down an entirely different path.

ജ • ൟ

Another method of altering the secondary successional path is by harvesting some or all of the overstory trees. Harvesting is the human activity that, like fire or wind or disease, alters the successional path of the forest. How the harvesting is carried out, which species are left in the overstory and understory, the composition of the soils, and many other factors influence the direction secondary succession will take following harvesting.

One of the most critical skills a good forester employs is an understanding of how the forest will respond to harvesting activities. With this understanding, the forester can influence harvesting activities to maximize the health, productivity, and diversity of the future forest. Knowledge of successional patterns of forest species and cover types is one key to the forester's success.

I pause again and look at the results of seventy years of running for the sky. Today a stunning northern red oak forest surrounds me in the winter light. But the forest floor is carpeted with small maple seedlings less than three feet tall. Unless something disturbs this site, it will be a maple forest in two hundred years, and the oak will be largely absent.

Succession in the Driftless Area

THE DRIFTLESS AREA: the name conjures a question. Driftless of what? A cloud drifts overhead. Driftwood washes up on shore. A drifter stands on the shoulders of the highway, his thumb pointing west. The Driftless Area—it must be a romantic place.

Here in Minnesota, this romantic image is grounded in a more geographical reality. "Drift" is a glacial term the geologists offer us. It refers to the debris that advancing glaciers churn up and carry on their cold backs as they slide down from the polar north. It is mineral soil and rock and gravel carried by the boiling ice on its slow, inexorable journey south. Driftless, or more mundanely, lacking a glacier, describes a pre-glacial landscape.

For some obscure reasons of topography, geology, or climate, the last great glacial advances of 14,000 to 8,000 years ago detoured around a certain portion of Wisconsin, Minnesota, and Iowa. This is now the driftless area. It is an older landscape. This hilly country in southeastern Minnesota, southwestern Wisconsin, and northeastern Iowa remains an echo of the past.

The landscape is furrowed with valleys and ridges, a miniature Appalachia on the northern great plains. Coulee Country, as it is known locally, is a channel geography, and all the channels lead to the Mississippi River. From the air, there is an almost magical change from ochre-hued agricultural fields to a green, crinkled landscape. Like veins in a leaf, these crinkles carry woodlands on their shoulders. Too steep to farm, these steeper slopes of the forest etch their way down to the Mississippi corridor.

꽁 • ꞔ

The Driftless Area is largely a neglected place. In the quiet valleys of the Root River, the Whitewater, the Turkey, the Rush, and the Coon, an interesting ecological succession is occurring in the forests. These wooded hills, now thick with oak, were once the hunting ground of the early American tribes. There were few forests here when early explorers first journeyed up the Mississippi. The hillsides were burned every spring to set back the land to grass and shrubs, forage for bison and deer.

With European settlement, though, the tribes moved west or disbanded, and the annual fires ceased. The oaks, well adapted to survive fire, with their root systems intact, resprouted with renewed vigor. Across the barren hills, northern red oaks, in stands up to ninety-five percent pure, lifted their branches to the sun and the sky. A hundred years later, magnificent oak forests cover the undulating slopes leading to the big river.

This is one of the ironies of European settlement. The same settlers who cut the great white pine forests and encouraged fire to rage to the north displaced the firemakers and discouraged fire in the Driftless Area. The settlers gave us back the red oak forests of southeastern Minnesota. And now, as we have become accustomed to these wonderful landscapes and the amenities they offer, nature is again playing tricks. The oaks are being replaced by other species.

Northern red oak (*Quercus rubra*) is not a long-lived species. Unlike its cousin, bur oak, which can survive over three hundred and fifty years, northern red oak starts to decline after about one hundred and thirty years. Rarely will the species live longer than one hundred and eighty years. While the exact process of decay is still not well understood, what is observed is usually heart rot.

Heart rot usually starts from a wound and moves to the center of the tree. At a certain point, the tree is no longer capable of supporting itself. A windstorm, which might sway the branches of a healthy tree, will suddenly split open one of these magnificent old oaks. On your next walk in the forest, it will be lying shattered across the path.

An examination of the tree butt will usually reveal a large hollow or rotted core with a ring of healthy wood forming the outer few inches. Sometimes only an inch or two of healthy cambium remains. Sooner or later the tree collapses.

Heart rot is not the only agent felling the oaks of the Driftless Area. Humans are at work, too. Red oak is the preferred species for much of the hardwood forest products industry. Red oak for kitchen cabinets. Red oak for tables and chairs. Red oak for hardwood floors. Red oak is a beautiful, strong and easily worked hardwood. In the early 1990s, over thirty percent of the ten billion board feet of hardwood lumber produced in the United States was red oak. The red oak forests of southeastern Minnesota were supplying millions of board feet of high quality lumber to world markets.

℠ • ℣

Some mourn the loss of the oaks, attributing it only to an overzealous lumber industry cutting oak for the flooring and furniture of the folks in the city. But nature has a more complex method than this.

The forest that is replacing the oaks is a more complex place. Ecologically it is more diverse. In the aging oak forest of the Driftless Area, where fire is gone, canopy gaps appear from dying trees. The gaps are the result of insects, disease, wind, and human hands. They let light into the forest floor. Maple, hickory and basswood, cherry, and elm, along with hack-

berry, boxelder, black walnut, white oak, and dogwood are all on the rise. These are the new and
expanding species of the area. They represent
increasing complexity and biological diversity. They
represent a more complex future.

We imagine that these oak forests have always
been in the Driftless Area. In fact, they are really
recent arrivals, like ourselves. I will miss the oak forest, but the oak will not be gone. It will be present in
the new forest, but not as the dominant species I
have known. It will be one of many species to reach
the light in the Driftless Area tomorrow.

Hay Creek

HAY CREEK SITS on the northern edge of the Driftless
Area, just outside Red Wing, Minnesota. It is a small
watershed that meanders down from the rich agricultural uplands, across the heavily wooded hillsides
and into a widening Mississippi River just north of
Lake Pepin. In a pastoral setting, Hay Creek is a gently flowing stream for most of its twenty-mile journey.
On one stretch, where it drops one hundred fifty feet
in two miles, it is a clear trout stream, with its shores
heavily wooded with oak.

Much of this hillside land was formerly owned
by farmers who used it to graze their cattle and cut

their fuel wood. In the late 1950s the state of Minnesota made a concerted effort to reduce sediment and erosion problems resulting from overzealous farming. This program began with the successful purchases of large tracts on the Whitewater River to the south, after serious floods had brought millions of tons of topsoil from the prairies cascading down the valley and into the Mississippi.

At Hay Creek the state found a few farmers willing to sell their land along the steepest portions of the creek. This area now comprises about 1,500 acres of the publicly owned Hay Creek Recreation Area.

As foresters and ecologists know well, red oak is a wonderful species, but it is difficult to regenerate. If you plant *Quercus rubra*, it will not immediately reach for the sky. It will dawdle near the ground while it puts down roots. Its neighbors, the aspens and boxelders and elm, will quickly pop out over it, and very few oak seedlings will make it into the light. This presents challenges if you wish to have oaks in the new forest.

In the past, fire took care of red oak by eliminating the competition. The maples, the ironwood, the hazel, and the birch were no match for the annual fires set by early tribes or that nature thunderbolted onto the prairie savanna. The tops of the young oaks were scorched by these fires, but their roots survived, and young, vigorous shoots of red oak sprang up and ascended into the canopy.

Red oak is also well adapted for fire when it reaches middle age. Its bark is thick and corky. When brush fires burn across the landscape of a fifty-year-old red oak forest, the bark will protect the community. The live cambial tissue is insulated from the heat by one inch of tough protective matting. The fire passes. The shrub layer and thin-skinned hardwoods are reduced in number, and the oaks continue their dominance of the site.

My interest in the puzzle of oak regeneration began at Esden lake. I had noticed that the red oak did not reappear in significant numbers after harvesting on my own land at Esden Lake. Even in a year of a good acorn crop, the seedlings would germinate, survive for a year or two, then wither and die.

In academia, immersed in the long literature on red oak, I learned that this lack of regeneration was a problem recognized by silviculturists across the country. The red oak forests of Pennsylvania, North Carolina, Wisconsin, Missouri, and other states were slowly disappearing. Lots of studies had been made, but reliable solutions were lacking. There were advocates of planting seedlings, but success in planting had been elusive. Too often the seedlings grew slowly and were over-topped by competing vegetation.

Other researchers advocated clear-cutting. This reduced the competition and encouraged stump-sprouting on the smaller oaks. The difficulty with clear-cutting was that most harvests were cutting oaks sixteen inches or larger and few of these stumps sprouted new shoots.

Finally, there were advocates of the shelterwood system of regeneration. Shelterwood involves leaving a portion of the canopy intact, as shelter, when regenerating a species. The partial shade of the residual trees protects the young seedlings from desiccation while limiting the growth of fast-growing pioneer species like birch and aspen. A shelterwood system, however, is difficult to implement. Furthermore, research results in the United States have been erratic.

With this in mind, I headed to the Driftless Area in search of a site to study red oak regeneration. I was in luck. A local state forester suggested that Hay Creek might fit the bill. He was establishing an experimental treatment there and was pleased that someone was interested in documenting the vegetation and changes in species composition.

When I arrived in the Hay Creek watershed for the first time in 1988, I fell in love with the undulating landscape and the intimate valley. It reminded me of a miniature Appalachian landscape. The beauty was intact, only the grandeur was lacking.

Hay Creek was a landscape of oaks. The hillsides of the northern and eastern slopes were packed with red oak, bur oak, and white oak. I pulled my truck to a halt and started up the trail. A small walnut plantation greeted me on the port side. It was healthy, a sign that the forester cared. Then the trail turned up the slope of the valley.

I climbed past an old orchard and began a slow ascent on the northern side of the valley. A sea of red oak arched their way over my head, and I spotted the experimental treatment area further up the trail to the west. Here, beneath the oaks, the ground cover was torn up. The ironwood saplings and hazel and buckthorn were flattened on the ground, leaving the mineral soil exposed. On three acres of the site, a bulldozer had passed recently. It had not touched the overstory oaks, but it had leveled the understory, pulling and scraping at the brush beneath the oaks. Torn roots of boxelder, buckthorn, and hazel lay drying in the dirt. To the untrained eye, this was destruction in paradise. To me, it was a bold experiment to regenerate oak.

An acorn fell by my side and onto the exposed mineral soil. The ground nearby was littered with acorns resting on the dark earth. Nearby, where the bulldozer had not passed, other acorns lay obscured and shrouded in the brush.

This was "site preparation" coincident with a good acorn crop before a harvest. Here was an attempt to mimic fire: to remove the competition before the oak seedlings were established and the seed source disappeared.

80 • CR

Over the next two and a half years, I spent many hours in the dirt on this ten-acre site. I brought hip chains and prisms to measure the area and the overstory. I brought yardsticks and nets to measure the crop of acorns. I brought a staff compass and put in seventy-four permanent plots on transects run out from metal stakes. On each of these plots I tallied the shrubs, the oaks, and the competing vegetation.

Nine months after starting, there was a harvest that removed the oak overstory. I needed a chainsaw to cut through the residual slash, find my plots and re-measure oak regeneration levels. I counted seedlings and seedling sprouts on the treated and untreated areas and the raspberries coming in a year later. I worked in the rain and the sun, ate many peanut butter sandwiches on the hillside, and filled pages of tally sheets with numbers. In spare moments, I sipped tea and watched the seasons change.

From time to time, other foresters and researchers would visit me on the site. My silviculture professor came down from the north and discovered a white pine plantation at the top of the hill, which interested him. My thesis advisor visited and tried to think of ways to expand the project for future funding. One day there was a field tour by the Society of American Foresters. Others were interested in the site and this spurred me on.

But most of the time I spent alone at Hay Creek, face to the ground, pencil and clipboard in hand. It was a special time, a time when the afternoon rains no longer diminished my smile. Seasons passed before my eyes, and I became intimate with the hillside. I listened to the landscape and watched it change.

Hay Creek was an initiation for me. I learned more than how to put in research plots and take field measurements. I learned the basics of research methodology and that nature is full of surprises. I learned that it was best not to have preconceived notions about hardwood regeneration.

Hay Creek was my training ground, and red oak was the species that taught me. It was time to return home to Esden Lake.

Autumnal Air:
The Bugs Are Down

MANY OF MY FRIENDS, particularly those from the city, make great protests about the mosquitoes at Esden Lake. They complain that one mosquito kept them up all night and glare at me for sympathy. But they won't find it. Mosquitoes are a fact of life in the north woods. It's best to adapt to them. After all, they outnumber us.

Perhaps the protestations go back to an old fear, to the times when mosquitoes killed our ancestors with malaria and other diseases. Perhaps my friends are used to controlled environments with air conditioning, skyways, and the like. Whatever the reason, it is a waste of energy to rail against mosquitoes. They were here long before we were, and they will be here long after we are gone. They are also a key protein in the food chain for the songbirds I love and the dragonflies I watch.

The mosquito hatch is an impressive display of the power of the small. In late May, with spring bursting around me, the mosquitoes return slowly. At first I notice them and smile. They are a sign of summer, of warmth, of songbirds in the forest. After a long winter, I welcome them. The early hatch is also a bit numb. The insects hover and buzz, but they are a little slow from their winter sleep.

A week later, however, I am no longer smiling. One evening, while planting my annual flower garden, these insects strike with fury. The air is calm and warm. The daylight is fading. Suddenly, in the midst of the pleasure of my garden, the wild hordes arrive and drive me onto my screened porch.

For two to three weeks in late June, when the mosquitoes are at their peak, I stay on my porch most of the time. Particularly at dusk, when the hum through the screening is fearsome, the mosquitoes own the outdoors landscape. I let them have their time.

But nature has her balancing mechanisms, and at Esden Lake the balance arrives with the dragonfly hatch. Remarkably timed to appear a few days after the mosquito swarm, one morning the driveway is layered by thousands of tiny black and white helicopter wings. Rising off the warm roadside in waves, the dragonflies are miracles of motion. When we invented the helicopter, we learned from the dragonfly. They can stop instantaneously in midair flight and hover motionlessly over a leaf. They can dart like hummingbirds through the tall grass, with their oversized black eyes perched on paper wings. They can make a ninety-degree turn and carve out a square in the air. Whence did nature collect this genus of creature?

The mosquitoes are no match for the agility and speed of these voracious wonders. Within two days, a noticeable decline in the daylight swarm occurs. Temporarily I can reclaim a walk to my mailbox.

Still, the mosquitoes that remain are a smarter breed. They seem aware of the dangers of daylight. They wait for the dusk to protect them from the eager eyes of the dragonflies. Now the bewitching hour follows the sun. My screened porch is a necessary retreat from the evening mob—a refuge from the welts. I sit in my rocking chair and listen to the music of millions of wings beyond the screen, and watch for the heat to pass. I am waiting to reenter the forest.

80 • ∞

It is now mid-August, and a cold front gives the air an edge it has lacked since May. Last night a front blew through Esden Lake. The high drama of wind and light against the sky is a precursor of the future. It is time for a walk in the woods, my first sojourn in six weeks.

For two months, the insects have owned the forest. Now I am going to test them to see if I can re-enter the forbidden zone. By early August, head well covered, I am game to try the trails again.

The deerflies are the principal targets of my investigation, for they are the most aggressive defenders of the faith in these parts. During their reign in July, they hover over my head like attack helicopters, their incessant hum turning even balanced minds toward madness. They bite through clothing, stick in my hair, dive bomb my eyes, and otherwise so disrupt my journey that I wonder what stupidity brought me to test them.

I enter the green jungle and immediately it is clear. The bugs are down. This is the surest sign of change in the seasons. There are other signs as well: plentiful signs amidst the wall of green growth around me. The lower branches on the sumac have started to yellow. The thistle is coming into bloom, its stunning violet orbs opening for the finches. Scattered aspen leaves litter the trail in umber and ochre. The fall seeders are ripening; ash branches droop low with clusters of samaras; chokecherries are now deep crimson, daring a bitter taste.

It is a treat to reenter the trail behind my home. A return after a long absence is sweet, as sweet as the first sound of the frogs proclaiming spring. In my absence, the forest has changed. In my absence, I have changed. With a new set of eyes and few bugs to distract them, I am ready for the future. Ready to go to work.

Regeneration and Release

The First Time

PLANTING TREES is an act of faith. Faith in the future. If it is your first time planting, you are in for a treat—and some hard work. Actually, the hard work comes first.

The first forest planting I organized was seven years before I became a forester. I ordered 1,500 trees from the state nursery. I then sent out invitations to my friends to join me for a tree-planting weekend over Memorial Day and prayed for good weather.

Today, fifteen hundred seedlings are no longer a large number of trees to plant. But fourteen years ago, when they arrived in the plastic and cardboard boxes, they felt overwhelming. Was I thinking that a group of baby-boomer friends from the city could capably plant over a thousand spruce and pine seedlings in three days?

Everyone arrived on Friday in high spirits, ready to plant and party. We set off to get a head start in good weather. By the end of the first day, three of us had put 300 in the ground, and already we could see that this was no trivial task to be finished in a couple of hours. Our arms were tired. The sod was thick. The rocks were abundant. The planting bars were heavy. I began to have doubts.

It turned out to be more work than romance to plant trees. First of all, the seedlings need to be kept moist. Dry roots equal death in planting. This means buckets, and water, and weight in the woods. Then there is the question of matching species to the site. If I turn my friends loose with seedlings and planting bars, I may find the seedlings planted anywhere. In our Memorial Day planting, this is what I found: red pine planted under the shade of maple and white spruce planted in the sand.

Finally, there is the planting tool itself—a heavy metal wedge, which requires significant exertion to create a pocket for the small seedling to call home. Lift, carry, stop . . . pause . . . raise the arms, thrust down into the earth (or rocks), pull back and forth, remove. It all sounds vaguely erotic until you try doing it fifty times in an hour.

There was more to planting than I had anticipated, "site selection," for starters. I had planned where half the seedlings would go. The remainder I was naively expecting to find homes along the edge of my road. Here, amongst the rocks and the raspberries and the sod, the competition was stiff. Would the seedlings survive?

Then there was the question of "site preparation." When you plant a garden, you don't merely plunk down your transplanted tomatoes or your carrot seed anywhere. You prepare a seed bed. I had prepared a one-acre patch of mineral soil turned up by a bulldozer. The remaining seedlings had to go in the ground without help. Sod, hazel, and ferns were their neighbors in the trenches. What kind of a chance did this give them?

My former partner T. was the patient and persistent one. Calmly she set out the second morning on her own, with another 200 seedlings in hand. I doubted her stamina, but by the day's end she had planted them all. I was impressed and grateful. Since that day I have learned that women frequently make better planters than men. There are exceptions, but women generally are better suited for the long, patient approach to work.

Eventually, everyone did plant some trees. Some of my friends stopped at fifteen, some at fifty. Once the novelty and romance had worn off and the reality of hard work set in, most opted for the lakeside porch, a beer, and good conversation. This was Memorial Day weekend, a time to party, not to work. A few hardy spirits each put one hundred seedlings in the ground, and two dear friends each planted more than two hundred. By Saturday night, though, I had a grumbling crew—and more than six hundred seedlings to go.

On Sunday my crew struck. Some of them started packing their bags for home. The others refused to budge from the porch as I flipped more pancakes and boiled more water for coffee. I didn't fight it.

What did I learn from this planting initiation? Urbanites love to talk trees, but don't expect your friends to plant on a holiday weekend. At least don't expect them to plant hundreds of seedlings. Perhaps thirty each would be a reasonable goal.

I also learned that I should have been a better instructor. Most folks don't know much about the care of tree seedlings. They let roots dry out on the ground. They dig holes that are too shallow or too deep. They force seedlings into holes with their roots in a J shape, sticking up. They don't pack the seedlings after planting, leaving air pockets around the roots. Planting trees is a more difficult art than it appears.

It was a good initiation for me. I was glad to see so many friends appear on the doorstep of my new home. And I didn't regret any part of the weekend when I headed out on Tuesday morning, planting bar and seedlings in hand. Sometimes it is best to do things yourself.

Experiences in Planting Failure

ELEVEN YEARS LATER, I stand on the site of the Memorial Day planting. The words of Bob Dylan echo softly:

> *There's no success like failure,*
> *and failure's no success at all.*

Dylan says it well, but it's a hard lesson to learn, whether in art or love or the business of planting trees.

In my first planting, over ninety percent of the fifteen hundred seedlings planted that Memorial Day weekend died within five years. Ninety-percent mortality is what we, as foresters, euphemistically call "plantation failure." Anything over fifty percent qualifies. I qualified with flying colors! What went wrong? Why did nine of ten seedlings perish in the ground?

A lot of things transpired which I was not prepared for. I lost ten to twenty percent in the first year due to poor planting techniques. Planting techniques include the manner in which the seedlings are cared for before they are put in the ground. Are the seedlings being stored in a cool dark place? Temperature is critical. If a box of seedlings sits in the sun for one hour, mortality will begin. The plastic bags, which wrap the seedlings and keep them moist will turn into ovens, baking their contents.

Planting techniques include the manner in which the seedlings are prepared for planting. The seedlings need to be separated and sorted. It is wise to remove the runts—seedlings with minimal root systems. They have a slim chance of survival in the real world. It is also important to trim the roots, which are longer than the planting hole is deep. This avoids root girdling and "J root," a common occurrence in open field plantings.

Proper planting technique includes mention of the type of planting equipment utilized since there are different tools to put trees in the ground. I now prefer to "auger" plant. It takes more time and is somewhat more costly, but with auger planting the seedling has its own loamy cylinder of soft soil to grow roots. This gives it a head start.

Another common planting tool is the hoedad. Professional planters in the western mountains have been using the hoedad with success for many years. It is a unique tool with the planting blade placed at approximately a ninety-degree angle to the handle. To employ it, you swing it like a pickax and create a cratered pocket, which the seedling nestles into. It takes time and practice to use a hoedad correctly, but it is time well spent.

There are two other conventional planting tools. One is the planting bar. A planting bar is exactly what it sounds like and is the most limited tool. It works well in light soils where there are no rocks or compacted soils. On rocky sites—sites with compacted soils or reforested sites—the planting bar is a poor choice of tools. A hoedad is preferred. Finally, on agricultural sites, there are machine planters, which plant trees quickly and come in a host of designs and sizes. In old agricultural fields or meadows, machine planting is an option worth consideration. But in reforested areas or on significant slopes, machine planting is usually not possible.

Today, I work with a professional crew. These folks have been planting trees for twenty years. Each year they plant more than a quarter of a million trees. They are led by a strong husband-and-wife team and are pleased to be able to share their knowledge and experience. They know how to work efficiently and effectively. Each member of the crew plants more than a thousand seedlings or transplants a day with an auger. They plant them correctly.

Healthy seedlings on a good site with proper soil preparation are merely a prelude to success. Poor planting techniques will cancel out all other factors. If the seedling is too shallow, its roots will stick out above the ground, dry out and die. If the seedling is too deep, the root collar will be buried, and moisture and oxygen will have a more difficult time reaching the roots. If the planting hole is not completely closed, air will dry out and kill the root system. There are fifty ways to plant a seedling poorly. A good crew knows this. In addition, their backs are strong.

Proper planting techniques make a happy childhood for trees. There will be substantial obstacles on the road to growth and survival. As an example, following our Memorial Day planting, nature conspired with a three-year spring and summer drought. Three years is a long time to be without adequate water. I made feeble efforts to carry water to some of these seedlings, but there was no way to water 1,500 seedlings. It was a poor year to plant—the worst drought in twenty years. Timing, they say, is everything for success. My timing was poor.

Drought is the single greatest cause for regeneration failure worldwide. One forest-products company in my area planted some of their sites three years in a row during this drought. They planted with improved stock, they planted with scarified soil, and they even planted with refrigerator trucks holding the seedlings. They had all the "right" tools, but they too had planting failure. After the third time, they gave up and sold the land.

ৡ • ও

Back at Esden Lake, about a third of the Memorial Day seedlings made it through the first three years, drought and all. However, there was a new problem awaiting them: competition. Before planting, I had cleared bare mineral soil to prepare the site. It looked like an agricultural field in spring. But mineral soil is an attractive seedbed for many native species. There were raspberries, thistle, and hazel for starters. And, for the main course, birch, aspen, and oak were looking for new territory to settle.

After three years, there was a wildflower patch to rival a magazine cover where the pine seedlings stood. There was thistle in glorious purple for the goldfinches. There was Indian paint-brush in brilliant cadmium, and raspberries for the bears. Also, there was birch over four feet tall. Meanwhile, the pine seedlings were still struggling at sixteen inches.

I considered spraying to save the remaining trees, but the Indian paintbrush was too beautiful to scold, and I like goldfinches too. Instead, I made half-hearted attempts to release the seedlings along the roadside with predictable results.

Now the pines had two strikes against them and were dwindling fast. The third strike came bounding out of the forest on four hooves. We know them as white-tailed deer.

Yes, the lovable north woods creatures, with shades of Bambi on their flanks, swooped through the woods in late winter in search of green food. The only green in town in late February, when diets are thin and browse thinner, were the buds of my dear pines. Mmmm good. The cloven hooves paused in the melting snow, and in a flash the seedlings were stripped bare. Needle-less stalks stood naked in the dwindling snow, victims of nature's hunger in late winter.

I was out and badly beaten. My plantation failure was approaching ninety percent. Left behind were a few stunted pine survivors and a few more spruce that the deer had chosen to leave behind. Such was the outcome of my first planting.

℘ • ℛ

Today, when landowners ask me if they should hire a planting crew to reforest their lands, I smile. Perhaps the question is a little like asking if a carpenter should frame up your house. Are you in good shape? Do you have the time? Are you willing to learn from your mistakes?

I'm not the only landowner who has lost more than fifty percent of the trees I put in the ground on my first try. Planting trees is as much of an art as a science; like all arts, it takes time and experience to learn how.

Success in planting is related to planning ahead. The species planted need to be matched to the site. If it is a sandy site, sugar maple will not survive. If wet, pine will do poorly. It is not the number of seedlings planted that counts; it is the number that survive. These little guys and gals are going to have their hands full out there in the big, bad world. There will be times of drought, raspberries crawling over them, and gophers needling at their feet. At least give them a fair start.

Natural Regeneration

FOR THIRTY YEARS Americans have been able to put men on the moon. That takes some doing, yet we still don't understand reproduction and regeneration of many tree species in the forest. We've been taking them for granted too long.

In the Lake States' forests, for example, a high percentage of the forests were harvested in the late 1800s and early 1900s. After the industry moved on, some lands were farmed. Many more acres burned and were abandoned. This is an old human habit that goes back hundreds of years. New Englanders did it in the 1700s, the British did it in the 1500s, and the Spanish did it in the 1300s.

If we glance further back in history we see the same pattern of abuse repeated many times. Early in the days of the Roman Empire, the Roman emperors had to send ships to Spain to transport the wood cut there so they could keep the hot baths going and the public pacified. The Greeks leveled their own forests. The Mesopotamians and Persians did the same thing four thousand years ago, and now, with a climatic change, there is a desert in their place. The French have an expression for this pattern of civilization: "*En avant, il y à le forêt . . . et après, le desert*" ("Before there is the forest, and afterwards the desert.")

In North America, we've been lucky so far. The trees have come back after our first try at deforestation. Quietly, without government subsidies, nature has reforested large portions of the North American continent on her own. How did she do it?

The causes of natural regeneration start with the past. What are the forces that shaped the look of the land? What story do the soils tell? Is this a sandy outwash plain from an old river bed, or is it glacial till formed as the boiling ice churned up the ground at its feet? Is it a loamy moraine built up where the ice paused or retreated? These questions offer clues as to the texture, the structure, and the moisture-holding capacity of the world beneath our feet.

The soils, it turns out, are far more important than many recognize. They determine what minerals the plants will find. They determine the amount of moisture held in the ground near the surface and the available organic material for growth. The three building blocks of soil science are clay, loam, and sand. These are categorized by particle size, sand being the largest and clay the smallest. With a high percentage of sand, water flows through the soil. The site becomes droughty, home to jack pine and blueberries. With a high percentage of clay, water backs up, and the site easily becomes saturated and short of oxygen, home to species like alder and black ash.

Soil conditions remind me of the story of the three bears. Is the porridge too hot, too cold or just right? Loam-sized soil particles are just the right size to hold some water and let some water pass through. Most plant species prefer their porridge just right.

Rainfall is another key to natural regeneration. Twenty-five inches of rain per annum is at the low end of forest tolerance in the temperate zone, below which only a few species live. Less than twenty-five inches of rain a year is roughly where the Great Plains begin.

Just as humans build their cities near rivers and oceans and lakes, trees build their forests on sites where water is plentiful. Outwash plains of sand don't hold much water. Here you'll find jack pine and blueberry and northern pin oak—natural regeneration adapted for droughty sites.

A few hundred yards from the outwash plain may lie an end moraine with loamy soils. Here a thick understory flourishes, with little light reaching the ground. Sugar maple, cherry, elm, red oak, butternut, dogwood, and leatherwood all grow well.

To regenerate naturally, some species need fire, and some need mineral soil, some need shade, and some need full sun. There are thousands of permutations and combinations to the variables. As the soils change, as the light changes, as the temperature changes, the species mixture changes.

Over ninety percent of the forests in the Lake States today have regenerated naturally. Humans have planted less than ten percent. Nature has learned how to grow trees on rocky ledges, on mining spoil sites. She has learned how to grow trees in the wetlands and on sand. She has learned how to grow trees following windstorms and fires. We can do worse than watch her and learn from her.

Release

AFTER PLANTING SEASON and the seedlings are in the ground, my arms are weary, my back is sore, and my boots are cracked and damp. In the bathtub as I soak my blistered feet, I recognize the labor of the past weeks. I will not see these trees full grown. I will not benefit from their shade or their nuts or their wood. And yet I know that I have planted for the future.

My blood is out there with those seedlings, and my impulse is to cling to them and their survival. But once the seedlings are in the ground, there is little that I can do to help them. They will have to adapt to animals nibbling at their feet, to winter winds and to summer drought. Some of them will not make it. I may be there to watch them and to encourage them, but I can not grow for them.

If I free the seedlings to their own lives, I free myself from attachment to their success. This is not an easy concept for me to accept. Most of my life is spent holding onto the past and to those I care for.

ঙ • ঙ

Recently my father was ill with a terminal condition. I reacted, predictably, by holding on. He was ready to move on, but, because I was not, I invented all sorts of reasons to cling to his presence. I called him frequently on the phone to hear his voice. I thought of excuses to visit him. I cried when I thought of his dying. Then it hit me. What would happen if I let go? Was it possible to love him and to let go of him at the same time?

My relationship with trees is similar. There may be a young stand of white pines growing in the half light in the forest. Over their heads is a canopy of hardwoods, aspen, oak, birch, and maple. What happens when the wind blows down one of these overstory oaks? The white pine, previously in the shadows, is released. The full sunlight over its head is capable of reaching it for the first time.

The pine responds. Slowly, at first, it reaches skyward toward the canopy. Growth increases year by year. Instead of putting on six inches of growth, it may put on two feet. This is natural succession for white pine. It is not a pioneer species designed for rapid early growth. It is a mid-successional species. It waits in the understory for its chance at the light. Sometimes its chance never arrives. If it grows under a long-lived oak, it may always remain a dwarf in stature. Meanwhile, the shorter-lived aspen has released its kin.

In his final weeks, my father reached a peaceful place. His eyes told me this. His eyes also asked me to accept the situation—to let go as he was. He told me that dying is a natural process. He had lived a good life. It was time for him to move on. This was his final lesson to his son. I was a slow learner. Finally, near the end, I understood. Once I understood, I, like the white pine, was free to thank him and to grow.

Shades of Gray

SILENTLY, NOVEMBER CREEPS across the landscape in the north country. Some recognize her arrival by the increasing hours of darkness. Others hear her coming when the rains turn to sleet, then ice. But I know November by her shades of gray.

I awaken in the morning and open the curtains to another somber sky. Time stands still. The shadows are dormant. A motionless whisper echoes in the bare branches.

Gray colors the landscape as well. The earth turns ashen. The bark on the oak mixes charcoal with umber. The sky softly cloaks its spreading gray palette over the earth. The world is going to sleep.

November offers release. It is time to let go of summer's shelter and fall's glory. It is time to breathe out all of the old emotions we carry on our backs, like so many wet leaves. Under my feet, the fading, fallen maple leaves pack into a mat of minerals and organic matter. Where is their glory now? I plod slowly across a somber landscape, kicking up past memories.

November is a time to simplify, a time to retreat to the warmth around the hearth. In the lengthening darkness, there are already new buds forming for next year's twigs. I notice them now that they are naked.

The air turns cold and reminds me of the fine line between life and death. The green and gold and crimson in the landscape disappear, and the darkness descends. If my firewood supply is plentiful and dry, November is the time to enjoy it. It is a time of reflection, a time to stir the soup over the stove, a time to invite friends for a hot meal. It is a time for the neglected side of life—time for mending clothes and writing letters. Now, as I waken to the ashes, I am grateful for the lessons November offers, grateful for the shades of gray.

Tales of Silviculture

The Ubiquitous Aspen

IN THE LAKE STATES FOREST, the story of the poplar family is a Cinderella tale. For years the family of poplars, and more specifically the species *Populus tremuloides* (quaking aspen) and *Populus grandidentata* (big tooth aspen) were considered the low-lifes of the northern forest, weed trees if you will. Exactly how their reputation emerged is not difficult to understand. In the early forests, poplars (also known as aspen) grew side by side with the majestic white pine, the long lived hemlock, the sturdy oak. These latter species all lived for hundreds of years, had considerable timber or tannin value and had a folklore of human history written on their limbs. There were treaty oaks and frigate-of-the-line white pines. Poor poplars, they came from the wrong side of the tracks.

The second chapter in this rags-to-riches story was written after the American Civil War. Until that time, paper had been made from rags. The Civil War created a huge demand for paper, resulting in a shortage of rags. To fuel the demand for paper in the 1860s, rags were imported all the way from Europe. This spurred the paper companies and chemists to experiment, and, in 1866, a Philadelphia company became the first to produce paper from poplars, using a lye solution to break down the fibers for pulping.

Over the years, this process has been improved and refined. Today, poplars are the building block for many of our quality papers: paper for magazines, for laser printers, for corporate reports. The ubiquitous aspen had found its market.

ᔓ • ᙚ

The silvicultural history of poplars is also an interesting tale. Poplars are generally short-lived pioneer species. They spring up following disturbances; in the Lake States, they followed the harvesting of the pineries. Poplars grow on dry, sandy sites with jack pine. They grow on the margins of wetlands with balsam fir and spruce and on loamy sites with maple and oak. They even invade the edges of the fields that the new northern farmers tried to clear. In short, poplars grow almost anywhere. They can take over the northern forest from their pedigree cousins in only a few years.

Early foresters responded to the cry against millions of acres of sprouting aspen by trying to convert sites to conifers, like red pine and white spruce. But poplars are tough to convert. They aren't interested in the missionary approach. Many sites where pines were planted quickly reverted to aspen.

Natural succession in the forest does not want to go backward, from hardwoods to conifers. Barring disturbances, such as fire, it wants to go forward, from softwoods to hardwoods. Attempts to push back the ecological clock usually end in failure, and the reforestation of cut-over aspen lands was no exception. Today, despite bulldozers and chemicals and chainsaws, many of these aspen sites have reverted to aspen.

Quaking aspen and big-toothed aspen have a couple of strong genetic codes. One strength, typical of sun-loving pioneer species, is their rapid early growth rate. As a family requiring full sun to thrive, poplars reach for the sky. It is not uncommon to find a three-year-old sprout that is ten feet tall. And, as the nursery folks have discovered, with a little genetic engineering, we can create a hybrid poplar with three times this growth rate. Poplars can be engineered for short term speed records.

Today the geneticists, industrial foresters and researchers are all touting poplars—hybrid poplars. They've learned how to cross-breed different strains of *Populus deltoides* (cottonwood) with strains of *Populus balsamea* (balm of gilead) and *Populus nigra* (European poplar). In the process, they've come up with some "super trees" that will grow ten feet in a year and be eight inches in diameter in ten years.

These trees are planted as cuttings in agricultural fields and are grown like a field crop. They are fertilized, sprayed with herbicides and cultivated. The short-term results are impressive, and for an industry in need of fiber to make more paper, hopes run high. Hybrid poplars represent the future for more than one paper company in America.

There are reasons, however, to be skeptical of the long-term picture for this miracle tree. Hybrids have significant drawbacks. Typically they are more susceptible to insect and disease outbreaks. In addition, there are unanswered questions about the soil conditions after a rotation or two of hybrid poplar. Will they deplete the calcium? Are high yields sustainable? There are other questions. Are these hybrids more susceptible to wind damage? No one yet knows the answers to these questions. For now, spirits are high in the agricultural camp. This is a new crop for beleaguered farmers and a new source of raw material for fiber. In the silvicultural world, the ubiquitous aspen has arrived.

Sugar Maple Blues

AT THE OTHER END of the ecological spectrum from aspen lies maple. Maples are marvelous trees. Some one hundred and fifty species are native to the north temperate zones of our planet. Hundreds more cultivars dot the yards of homeowners worldwide.

Of all the maples, sugar maple (*Acer saccarum*) is the most celebrated here in the United States. A thousand years ago, tribal cultures valued its boiled sap as one of the earliest food sweeteners. Three hundred years ago, furniture makers appreciated its unique grain patterns, its durability, and its color. Today, thousands crowd the highways to see its brilliant colors in the autumn.

With such an illustrious history, it is surprising to learn another side to maple's story. Listen in to a group of foresters in the woods and you might catch disparaging comments. Watch farmers along field sides with chainsaws and chemicals battling another member of the maple family, the boxelder. Ask wildlife managers about their concerns as maples invade old oak forests. What is all the fuss about?

Today, sugar maple and its cousins red maple and boxelder are increasing in dominance over large portions of the northern hardwood forest. As shade-tolerant species, maples arrive quietly. Born by winged seeds and warm-blooded vertebrates, the maples land unnoticed under the shade of the oaks, aspens, birches, and pines. It is common to see a literal carpet of sugar maple seedlings beneath an aging red oak canopy. Given time, the seedlings grow into saplings and the ecosystem beneath them changes dramatically. With heavy, thick leaves, the maples shade out the seedlings of red oak, white oak, birch, red pine, hickory, and cherry. Many tree species of great value to humans, forest mammals, and birds are crowded out of the new maple forest.

Historically, maple forests occurred where fires were rare. In Minnesota, there was a region known as the Big Woods. The Big Woods got its name from the settlers moving west in the nineteenth century. After miles of scrub oak and pine savannas, they ventured into a heavy forest of maple and oak. They called it the Big Woods.

Early ecologist Rexford Daubenmire studied these woods to try to determine their origin. What Daubenmire discovered and wrote about in his groundbreaking work, *The Big Woods of Minnesota*, was forest succession. He was able to show that there was a high correlation between the abundance of mature maple in the Big Woods and the frequency of fire. It turned out that the Big Woods was protected from frequent prairie fires by a series of rivers and lakes to the west, south, and east. It was an ideal spot for maple.

Daubenmire documented that sugar maple needs a protected, fire-free environment to flourish. It prefers moist sites, where water is plentiful, and it likes undisturbed oak forests, where partial shade keeps down the competition. In short, it likes peace and quiet. If you wish to find maple, look for it in protected, fire-free places.

Today, with the help of Smoky the Bear, we have removed maple's great enemy, fire, from the forested landscape. Without fire the new forests are missing many of their old friends. The oak forests, for example, are disappearing from part of the northern landscape, particularly on fertile sites. In their place, the maple slowly rises. Ecological succession is changing the forest landscape, and we have had a major hand in this process. For those who prefer the oak forests of the past, this is the sugar maple blues.

Red Pine Maligned

RED PINE (*PINUS RESINOSA*) has a poor reputation in my neighborhood. Foresters plant red pine in plantations and short-sighted folks call them "ecological deserts" and "monoculture wastelands." When I hear these critiques, I wait for the venom to pass, then invite the critics for a walk.

Up in Cass County is a small, red pine plantation I have been tending recently. It was planted in 1964 on poor agricultural land. The soils are light and sandy—perfect for pine. The planting was typical for the time—very close spacing with excellent survival rates.

The result of this work was a jam-packed jar of pine-spindly pine poles with a carpet of needles. Seven years ago it was difficult to walk through this darkened place. Dead pine branches barred paths in every direction. It was eerie and still and dark. The pines let in almost no sound and almost no light—like purgatory at twilight. There were no shrubs or herbs.

Then I thinned it. We removed about forty percent of the trees in an operation called a "systematic" thinning. Whole rows of trees were removed, leaving space in the middle for the others to grow. We hauled everything out of there, sent the poles off to the pulp mill to make the paper you're holding, and shipped the bolts to the local stud mill for two-by-fours. The tops were piled and burned to limit bark beetle infestation. The result was a clean-looking silvicultural treatment and an easier place to walk.

That was seven years ago. Now I don't recognize the place. A green band of herbs and forbs blankets the ground where before there were only needles. There is clover, hog-peanut, aster. In the spring the wildflowers are glorious—hepatica, bluebells, and daisies. Raspberries and hazel are coming in where the sunlight is strongest. Birds chirp where there used to be silence. The forest is alive.

The pines are looking better, too. They are slowly filling out in a graceful way, tall soldiers bending in the light. The smell of pine pitch fills my nostrils. The sun filters through the half canopy. The sound of the wind, a distant roar, a mysterious whisper in the needled leaves, makes me pause.

This is not a wasteland or a desert. This is a healthy forest. It is a fine place for a morning walk, a fine place for deer to browse, a fine place for birds and small mammals to make their homes.

Too many people hold an oversimplified view of this healthy plantation—a view that growing trees for forest products is an unhealthy practice. By this standard, all forests should be left alone for nature to handle. I often wonder where the food on our table would come from if we had the same view of agriculture.

It seems to me that growing healthy forest plantations is one step in the right direction. These plantations not only provide essential products; they also concentrate the resource base on a relatively small number of acres. Our fiber needs can thus be met more efficiently, making other woodland acres available for different benefits, such as recreational activities and old-growth ecosystems.

It is true that there is a phase in a plantation cycle when very little besides the species of choice grows there. But is this a problem? Glance at the millions of acres of corn along the highway, or the lettuce fields in California, the wheat fields in Kansas, the potato fields in Minnesota, the tomato fields in Florida. Do you have difficulty with these ecological deserts? Why not? All of these are much more intensively grown than any forest plantation I know, and they also have the added problem of chemical applications to the soil and plants.

The truth is that, ecologically, red pine plantations are a natural step forward in succession from the agricultural fields on which they were planted. Red pine, it turns out, is an early successional species, a pioneer. It needs full sunlight and limited competition to thrive in the early years. These are precisely the characteristics present after forest fires in the pre-settlement forests. The pines seeded in the ashes and bare mineral soil and grew quickly in the open light. Today we have removed forest fires from most of the landscape. How else will mother nature regenerate this species except by planting?

The old agricultural fields of the north woods mimic the needs for red pine regeneration better than any other ecosystem without fire. Open light and low vegetative competition are present in an old agricultural field once the sod layer has been removed. This explains why so many of the red pine plantings on old farms have succeeded. The disturbance that prepares the site is farming, not fire. Both types of disturbances prepare sites on light soils for pine.

Nature's funny this way. We see a landscape in one state of being and quickly pass judgment on it. We need to learn to be patient, for nature is full of surprises, such as the red oak seedlings that now pepper the ground of my red pine plantation. Where did they come from?

I look around as I stand in the shadows of the pines. There isn't a mature red oak within my eyesight in any direction. It's all pine to the north for a quarter mile, aspen to the west and south, birch to the east. And yet here are these red oak seedlings. They're unmistakable. I pause and watch the blue jay fly by. A squirrel chatters from a nearby limb. Do they know something that I don't know?

As foresters, we've had our share of failures. Red pine plantations, however, are one of our success stories.

White Pine: Death and Rebirth

THERE IS NO OTHER SPECIES in eastern North America to which Americans owe so much. Other species, like oak, may vie for our historical heritage, and still others, like chestnut, have been decimated more severely. But no other North American species can approach the mythological *Pinus strobus* when it comes to persecution and rebirth.

Standing erect and tall, head and shoulders above all the others in the eastern forest, white pine was our sentinel. A beacon in the time of darkness, we relied on the pine to build our homes, to shelter our tired limbs, to protect us from the cold north wind. White pine was majestic, graceful and serene. We counted on its presence in the forest and under our feet.

Here is a tree that was fought over before the American Revolution, when the British navy coveted its tall straight boles for its men-o-war. Here is a tree that built the industrial cities and towns of nineteenth-century North America. Here is a tree that was cut in Maine, in Massachusetts, in Michigan, in Minnesota.

After all white pine had given us, brave pioneers of the western front, we tried to kill it. We burned up its offspring with the flammable debris we left behind after harvesting. Holocaust fires followed. We plowed it under on the northern front for farmland and pasture. And then we accidentally imported a foreign disease, blister rust, to finish it off.

When early foresters were trying to replant white pine seedlings in the northeast and Lake States in the early twentieth century, they ran short of nursery stock. So white pine seedlings grown in Europe were imported to satisfy the demand. Unbeknownst to the foresters, the pathologists, or the nursery growers, some of these European seedlings carried the spore of a European fungus, *Cronartium ribicola.* We know it today as blister rust. It is easy to spot because of the white-orange blister-like spores that appear on white pine branches in the late stages of the disease cycle. Because it is an introduced disease, our native pines have little resistance to the fungus, which girdles and kills the cambial tissue and eventually kills the infected tree.

Blister rust spreads, as do most fungi, by airborne spores. Hence, it is difficult to detect and treat. By the time pathologists understood the disease, we had almost succeeded in wiping out the white pines in Minnesota. Fortunately, nature is resilient and a few have survived in the forest.

The lumbermen missed some of these trees. Others survived the blister rust invasion. Isolated lone sentinels, too young or too limby or too remote to cut, these survivors grew and produced seed. The winds arrived and carried the seed to nearby fields and highway right-of-ways. Today, white pine is making a comeback in eastern North America.

When farmers abandoned the fields they had plowed on the stony soils of New England, white pine moved back. It reestablished itself in the grass and brush. Slowly, in clusters like lily pads, the seedlings took hold. They followed America west again one hundred years after the saw-timber trail. Like a ghost of Christmas past, white pine popped up to give us hope for the future.

Today there are white pine forests in the canopies of Massachusetts and Maine. These forests appear old, but we know better. They are pioneers following abandonment of farming fields in the nineteenth century. The stone walls near their feet attest to the agricultural history that bred them.

Will we take care of white pine this time? It's too early to tell. This species helped give us a golden past. It offers us the potential of a prosperous future. This time it needs us on its side.

Crop Tree Release

ALL OF US have favored spots. Some of these havens may be small —a corner of a sofa, a view from a window. Others are less intimate—a winding country highway, a bridge over a bay, a soft-sanded beach on the ocean. For me, one of these special places is a small oak hillside I tend in the north woods of Minnesota. This grove of northern red oaks reminds me of my time at Hay Creek.

As with any location that is visited over time, the landscape changes with the seasons. In winter, on snowshoes, it is silent and white. When I pause, there is a stasis in the air, and only the wind whispers in response. Like a snapshot, it freezes time. Only a few mammal tracks hint at life beneath the surface. Frozen, immobile, dark gray cylinders rise off their white foundation—the oaks, a thousand sentinels to the sky. It is peaceful but cold, and I do not linger.

In the spring and summer it is a different place. Scarlet tanagers and reclusive warblers, travelers from exotic forests to the south, glide in its shadows, their voices carrying to the hills beyond. Easter-white trilliums appear in congregations, with thousands of three-lobed petals open to the sunlight. The fragrance of honeysuckle passes with the wind.

In the autumn, the air chills and the light under the canopy turns gold. There are sharp sounds under my feet as I tread the freshly fallen leaves. The dark trunks of the oaks stand out against the orange and the red of the maples. A sprinkle of acorns, recent arrivals from above, lie scattered at my feet. When I kneel, I note that their hard shells have dented the mineral soil on the forest floor. Occasionally, as I pause, a breeze of wind brings a fresh release from the sky. One of them strikes a branch nearby with a surprisingly sharp echo, then ricochets into the distance.

The oaks on the hillside are beautiful, straight-trunked trees rising thirty or forty feet before they branch. They will make clear-grained floors, tables, and moldings someday. Now, at age seventy, the trees are still young. They have seventy years ahead of them in the forest, maybe more if I treat them right.

On any plot in the forest may be one or two of these magnificent warriors standing alongside some less able companions. Their companions may have lost a limb early in life. They may have gotten a slow start seventy years ago. They may have grown up beside a large boulder. For any of a thousand reasons, they are weaker than their neighbors and show their weakness with small diameters and thin crowns.

From my point of view, these "cull" trees limit the growth of the others. Their branches are filling the space where canopies spread. If I remove about a third of them, the rest of the forest will be stronger and healthier. It's like thinning my lettuce patch. This is a simple concept, but with a seventy-year-old oak stand it's a challenge to implement this carefully. These are not five-pound corn stalks I am talking about removing but two-ton giants that have been rooted to the same spot for decades. Their branches intermingle and arch over a hillside. They do not want to move.

I ask a local logger to enter this undisturbed red oak site. The trees are located on a hillside with thirty- to forty-degree slopes. I ask him to fell, limb and haul the marked trees without damaging the remaining crop trees. I want this accomplished in an efficient and safe manner. As a bonus, I ask him to do this in December, when the ground is protected by two feet of snow, and the temperature is minus ten degrees Fahrenheit in the morning. This may not sound like a kind request, but it is the best method for improving the quality of my woodlot without causing undue damage to the soil or the site. During the dormant season, the oaks are sleeping and the hillside is protected with snow.

Crop tree release is one silvicultural method of improving the health and productivity of my oak forest. The crop trees are given more light, more moisture, and more growing room. This promotes their health. Their crowns grow larger and produce more acorns for the animals at their feet. Their roots systems grow stronger and are better able to withstand dry years. Their foliage grows thicker and produces more starches and sugars for warding off disease. As the oaks grow larger, a hike on the trail beneath them brings greater pleasure and satisfaction. I pause beneath one of the spreading oaks. All of us benefit from the forester's active hand in silvicultural practices.

Paul Bunyan's World

Pine Musings

I STAND IN THE SHADOW of a large pine tree near Esden Lake. This one offers me an opportunity to be impressed. Quietly growing for more than 100 years, this white pine has developed a presence felt by those who pass nearby. In a sense, the tree has created its own world under its crown. Squirrels, porcupines, and owls call its branches home. A young sugar maple grows in its shade.

I enter this quiet world and sit in its shadows. I enter to observe the majesty of the tree and its tall stature. I also enter to give it another life, a life on my walls, under my feet, in front of my reading lamp. Yes, I am speaking of harvesting.

Humans have been harvesting trees for a long time, probably longer than they have been harvesting agricultural crops. Civilizations have been built on trees. Mesopotamia and the Fertile Crescent were once wooded. Crete was wooded. Pre-Hellenic Greece and Italy were heavily wooded in the early days of the Roman republic. Germany was a primeval forest, as were England and eastern North America. In all of these areas, civilizations rose and utilized their forests. Humans erected houses, built ships to cross oceans, burned fuel wood to cook and to stay warm, and made paper to communicate with others. And this does not

begin to tell the story of the clean water the forest filters or the oxygen it creates. In short, we need the forest and have depended on it for thousands of years.

In North America, we have been spoiled. Nature has provided us with seemingly endless new
forests to use. At first the eastern forest was bound-
less, then it was tamed and cut. The Lake States' forest was next. Now in the Pacific Northwest we are following the same story.

But nature's bounty is limited. The Mesopotamians found this out after they had cut almost all
of their forests three thousand years ago. The Ro-
mans learned the same lesson after they had denud-
ed Italy two thousand years ago. The Spanish repeated the same process less than a thousand years ago
with their great galleons and a hunger for wealth.
Today, much of Spain is arid and treeless.

If we cut down a forest once, a new forest will
usually return. If we cut it again and again, add cattle for pasture, and then throw in soil erosion and climatic change, the most likely result will be a desert.
To repeat: "*En avant il y a le forêt . . . et après le
desert.*"

Thus far we have been lucky in North America.
In the twentieth century, oil has given us a second
chance. Around the turn of the century, in 1910, we
were well on our way to a timber famine. Harvesting in
the country outstripped growth by a factor of two to one.
Those were the days when the steamboats and the rail-

roads were fueled with wood and when most homes heated with wood. Oil changed all this. Rather than using hydrocarbons stored above ground, Americans learned how to use hydrocarbons stored below the surface of the planet, hydrocarbons that were plants and forests hundreds of millions of years ago.

This was an ingenious step. With the discovery and use of oil, pressure on our forests diminished. Our transportation system converted largely to oil. Oil heated our homes, and the demand for wood decreased substantially. The eastern forests began to return. As transportation efficiency improved, the fertile prairie soils of the midwest became the breadbasket of the country. Old agricultural fields in New England regenerated to oak and birch and pine. The forests of the Lake States succeeded to aspen and oak. In 1980, despite population growth and suburban development, there were five million more acres of productive forestlands in the United States than there had been in 1920.

How long will this trend continue or has it already begun to reverse? How long will we be able to dart about in hydrocarbon-fueled vehicles while preaching about protecting our forests? The petrochemical resource is finite. Sooner or later, we will need to return to our forests for hydrocarbons, as witnessed by the increased use of fuelwood following the 1973 Arab oil embargo. It is fine to talk about protecting rain forests and owls as we jet at 35,000 feet, leaving burned hydrocarbons in our wake, but will we be prepared to sacrifice our consumptive ways when oil is less plentiful and our children need homes? Have we learned anything from history? Will we be prepared for the future?

I am still seated under the same white pine. My mind has journeyed across time, and the wind brings it back to the present. This white pine has given me something. It is healthy for me to return the favor, and it is in my interest to do so. I will need to plant more trees. With knowledge of the past and patience for the future, I can provide the same paper and tables and hardwood floors for our great-grandchildren. Anything less is too little.

Tree Felling— The Seven-Fold Path

MANY WORDS have been written about felling trees, useful words about safety and equipment and technique. You probably have heard them so we will talk instead about philosophy.

The Buddha would probably object to the metaphor used in connection to felling trees, but with all due respect, there are similarities of approach and action: right thought, right attitude, right action.

I. Planning: Respecting the Forest

First, there should be a plan. What is the purpose of my excursion with the chainsaw? What area of the forest needs attention? I must decide which trees to fell—or perhaps it is better to speak of which trees to leave and why. Cull trees, release trees, improvement cuts, crown thinning for crop trees: these are all terms foresters use for certain kinds of trees in the forest. I am not merely gathering fuel wood, I am creating a forest for the future.

Cull trees are the trees I wish to remove from the forest. They are leaning, forked, dying or blocking the light of a healthy neighbor. They are generally in poor health or have been wounded by a prior adversity.

Release trees, or crop trees, are just the opposite. Those are the ones I wish to keep. They are healthy, straight and beautiful. The usual procedure is to locate the crop tree first, and then look for the cull trees around it. This is the planning process. If I don't have a plan, I save my chainsaw for another day.

There is a good reason to take my time here. Once that tree is on the ground, no amount of prayer will put it back in an upright position. Respecting the forest means that I consider not merely my immediate needs—firewood for winter, a better view of the mountainside, sawlogs for the mill—but also the long term. What will this forest be like after this tree is gone?

What will it be like in fifty years? Am I making the best choice here? Am I thinking about the health of the forest in the future as well as about the product I am harvesting? The right way respects the forest.

II. Preparing the Mind and Body

Safety is the foremost task of the day. Am I fresh and ready? Is my mind focused on the task at hand? Do I have the right equipment? Is it in proper working order?

Here I prepare myself with right mind. With right mind, the process is safe and complete. Without it, I face the danger of the unexpected. Is my saw sharpened? Am I wearing chaps, goggles, and a hard hat? My safety is my responsibility. So is my mental attitude. I take them seriously. A moment of preparation may save my life. There is no rush.

III. Respecting the Tree

As I plan the harvest, I try to consider the individual trees to be removed. Which direction does the tree want to fall? Many hours have been spent with hung up trees, pinched chainsaws, and even injuries due to haste here. Is the tree a leaner? Is it a windy day? Is the bole rotten? I stand under it. Which way does it want to fall? How much leeway in its trajectory do I have? One tree hung up on its neighbor may cost me half a day. Worse, it may cost injury. If I am in doubt, I leave the tree alone or let a professional handle it. I don't have to fell every tree in my plan. I can be flexible.

IV. Preparing the Site

Now it is time to remove the nearby brush and limbs that hinder my work. This is more important than it may seem. Brush and limbs in my path and in the path of the falling tree mean that unexpected things can happen. I do not want unexpected things to happen. I want to be in control of the whole process.

Without a clean area around the base of the tree, I could get careless with my notch cut. The notch cut or undercut is the most important cut I will make as it determines the direction and manner in which the tree will fall. This is not a place for short-cuts. It takes time. I plan for that escape route to be clean and clear. I look at the top of the tree one more time to certify its health, position, and lean. Then I notch the tree carefully. I do not make the notch more than one-quarter of the diameter deep, following the technical manual that came with the saw. I pause and think again after the notch is made. Is it the proper depth? What does the cambium look like? In all of this, I respect the tree.

V. The Dramatic Moment

The undercut is not deep. It is only a notch. But it is more than a slice of pie. The undercut will determine the location of the fall. It tells me the quality of the tree. Is it solid toward the center? Is there a flair on one side?

Now I approach the tree one last time in its standing position. This is the time to be focused. Is my escape route clear? Are there neighboring trees in the direction of its fall? If everything meets my satisfaction, and my guts tell me that I'm ready, it's time to finish the task.

The chain roars into action. The chips spit back. I glance skyward. I prepare for the back-cut. I will leave a hinge. I will finish the cut about an inch above the undercut and not flush with it. This is the moment the adrenaline rises. The eyes are skyward too, watching that top. Slow and steady, the chips fly. The eyes glance skyward, then to the trunk, then skyward. Some professional loggers keep their eyes on the top once the saw is firmly into the back-cut. They are looking for the first signs of shudder and movement.

This is the moment of drama for which I have prepared. There need be no frantic action. Felling a tree is an act I treat with respect. Finally, the top quivers, and its bare motion begins. I pull the saw and flick it off. I step back quickly with my eyes on my escape route. I do not linger near the tree. Objects coming from the sky and being launched from the forest floor can make life too interesting. Heads up, Peter.

Down, down, down, it swings, gathering momentum as it plunges earthward. The ground trembles with the sound and the fury of the giant felled. Then all is strangely still. My heart is still pounding.

VI. Examining History

I wait for the ground to settle. I watch for loose things in the branches above me that may be released by the disturbance. Slowly I approach the felled tree. I am a bit in awe of what has occurred. The forest has changed forever. New light reaches the forest floor. I glance up, and there is a new canopy gap in the forest.

Did the tree come down where I had reserved a spot for it? Why not? To fell a tree and have it rest where planned—that is control. It is a goal toward better and safer woods work. If the tree is not where I expected it to be, I try to understand why.

I may wish to examine the stump, which offers a history lesson. Its age, health, and family history are all written here in the rings. The tree offers a view of its past, and understanding it will help me understand my woodland.

Perhaps I will pause, relax and look at the beauty of the cross-grained wood. I breathe in the odor of the fresh chips at my feet and think about other stumps left in the forest. I glance at the new canopy gap. There is light where there was shade. There is satisfaction in my eyes. There is respect in my heart.

VII. Cleaning Up and Moving On

All processes need completion. After the high drama, there is still much to do. The limbing, the bucking into lengths, the hauling, and the stacking are all hard work. It can take fifteen minutes to fell a tree. It can take a day to cut it up and haul it away.

I go to work to utilize the products this tree has to offer. It may offer flooring for my home, siding for my woodshed, or heat for winter cold. It may become a log for the male grouse nearby to drum, a shelter for the neighborhood squirrel, or food for wintering deer. In a sense, the wood's life has just begun.

Lake States Woodsmen Today

MANY STORIES have been told about the early pineries in the Lake States and the men who tamed those forests. Most of our midwestern cities were built with those pines, and you can still see the massive twelve-inch beams as you walk the halls of the warehouse districts of Minneapolis, St. Louis, or Kansas City.

The men who worked in the forest in the late nineteenth century were, for the most part, recent immigrants from Europe. Woods work did not pay well, and the life was hard. Their bosses were migrants from the pineries of Maine and New York. They knew their pine, and, with recent inventions like the railroad, they knew how to move it to growing markets.

These men knew much less about regenerating the pine. In many places, new settlers were pleased to purchase the cut-over land after the loggers had passed and turn it into farmland. Anyway, there was always more "virgin timber" to the west.

When the pineries in the Lake States were mined out, many of the biggest names, such as Frederick Weyerhaeuser, who made his first fortune in Wisconsin and Minnesota, moved on to the northwest to grow their mills into empires. Others, such as Charles Walker, for whom a town in northern Minnesota had already been named, chose to invest their new wealth in other ways. Walker, made wealthy by white pine, founded and funded the Walker Art Center in Minneapolis. He went from timber baron to art philanthropist in one generation.

As the twenty-first century approaches, the men who work in the forest are not much different from their predecessors. Even today, these men are a special breed. Stand in the shadow of a large pine tree and you will understand why. The trees infect the men with their presence. Alone, chainsaw in hand, the logger feels the honor of the evergreen above him. He will pause a moment, look at the tree, and then go to work. And work it is, dangerous, difficult work, when done by hand.

I take the time to observe a skilled woodsman. I note the dignity and the grace to his actions. Without haste or waste he prepares, notches and fells the trees.

Delimbing is efficient and quick. His saw is kept razor sharp. The cables are run out with ease. When a skilled woodsman works on a tree, whether felling or limbing or bucking, he approaches it with respect and a minimum of movement. He knows each species by its personality—which ones are prone to hollow centers, or tough limbs or split tops. He works in the rain, in the snow and in the cold.

To watch a man fell one of these large old pines is memorable. The notch is cut quickly, deliberately, to the right depth and in the right direction. The sawdust flies. The back cut is made with confidence and intensity. At a certain time, all eyes turn upward. Even as he cuts, the feller glances skyward with the neck of an owl. When will the pine start to quiver? When will it inch forward?

ॐ • ॐ

In my mind, the great tragedy of woods work today is not the loss of the trees, for they will return sooner than we think. It is the loss of hand work by these men of Paul Bunyan stature. Large machines are now harvesting many woodlots in the United States. These machines walk up to a tree, sever it, delimb it and carry it to a nearby post-mortem pile. The machines distance the men from the tree. The machine is never silent. It growls and moves on. The man grows more distant from the trees he fells. He may no longer touch them or stand at their feet or straddle their limbs. The new woods worker does not even carry a saw. He is buckled into a three-hundred-horse-power machine and surrounded by metal in the forest.

How does this affect the harvest? Instead of being a hunter who takes his own animal, the new woods worker works in a slaughterhouse, where production quotas are the guideline and thousands of dead corpses pass through the machine every day. The trees become known by how many "sticks" they make and how quickly they can be processed. The old woodsman working alone or with a partner was satisfied with ten cords of production a day. The new woods worker needs five times that to pay for the equipment he operates and for his larger crew.

The forests today show the results of these changes. There are larger clear-cuts today than fifty years ago, partially because machinery on tracks operates more efficiently. It is expensive to move large equipment from site to site. There is the risk of more soil compaction because these machines weigh more than a skidder or a man with a chainsaw. No one knows the long-term effects of the changes in harvesting techniques, but the days of the old woodsman are passing.

Perhaps the new woods workers will be more educated. Perhaps they will know from computer controls which trees in the forest are worth saving for the future. Perhaps they will learn to love and respect the forest as they sit in an air-conditioned cab. Something, however, will have been lost. The myth of Paul Bunyan is real in the north woods. It is a myth worth remembering and honoring.

Wood Market Mumbo Jumbo

WOOD PRODUCTS are commodities: boards, chips, and pulp. When I think of commodities, I normally think of posted market prices—of quotes on the financial page—of an understandable price for an understandable product. But the fiber on my wood lot doesn't fit into this mold.

When I try to pin down "stumpage" prices for the pine, the oak, and the aspen growing on my woodlot, I am quickly confounded by a myriad of confusing data and terminology. One logger offers me a price per ton for a pine thinning, another a price per cord. A third logger offers me two prices, one for bolts and one for pulpwood. I am getting confused. One outfit will scale the high-quality trees in Doyle scale; another will make an offer in Scribner scale; and the local forester may quote me in International ¼-inch scale.

How do I make sense of all of this? Lost in the mumbo-jumbo terms, I may end up trusting the logger to tell me how much wood is coming out of the woodlot. Is this my best choice? I remember the story of the fox guarding the hen house.

As a forester, I sometimes receive calls from landowners who are in over their heads. They've agreed to let a logger harvest their land, and they slowly find themselves losing control. The price, which seemed fair a month earlier, now feels low. The quantity of wood coming off their woodlot is greater than they expected, and there is damage in the residual forest stand.

These landowners have decided to sail their own ship out to sea; now a storm has come up, and they're asking me for help. Why didn't they call me earlier? Is it because they are trying to save a few dollars? Is it because they are proud? Is it because they fear someone suggesting what to do on their own land?

I don't have the answers to these questions, but I do know that landowners usually suffer without professional help. A forester knows what is out there in the woods and how to measure it. He or she knows what the market terms mean and what current market conditions are. A forester is on the landowner's side.

As a landowner, it is in my best interest to learn as much as possible about the way the wood markets work. What qualifies as a sawlog? What qualifies as pulp? Who is buying oak for their sawmill down the road? Who is buying aspen for their contract with the paper company thirty miles away? What does the grading scale in hardwood sawtimber mean?

There is a reason for much of the mumbo jumbo of market terms. The stumpage in my woods has no set market price because many factors influence its value.

The most important of these fall into three categories: 1) What is the quality of the trees? 2) What is the quantity of wood I wish to sell? 3) What is the status of local markets? The more I learn about these variables, the stronger will become my bargaining position.

All trees are not created equal. Some are small or crooked and will make only fuel wood. Some are small but straight and of the right species (like aspen or spruce) to make pulp but not boards. Some are straight and large and will make hardwood or softwood boards. Some grow in clumps and are easy to harvest. Others grow scattered about on a steep slope. Some grow near a road or a woods trail. Others grow across a bog or a wetland and are difficult to harvest. Multiply these variables by topography, soil type, season of harvest and type of harvesting equipment, and there are the makings of a complex stew. The next time you ask a logger for the stumpage price of aspen, you won't be surprised if he responds, "It all depends. . . ." He's telling the truth.

In my neighborhood, there are three Fortune 500 companies that buy aspen. They use it as raw material to feed their modern, multi-million-dollar mills. As a result, aspen prices tend to be stable and strong. On the other hand, two of the local pine sawmills have shut down in the past five years so pine is more difficult to market today.

Finally, there is the question of measures. Learning to estimate the volumes of aspen and oak and pine in the woods takes time, experience, and skill. There are no shortcuts for this. If you think that you have 100 cords of wood and someone offers you $2,000 for it, you may be happy. If it turns out you had 150 cords of wood, you may feel differently. This experience is not rare. It happens to landowners hundreds of times a year.

By the same token, on small diameter wood, the differences between the Doyle, Scribner and International ¼-inch scale can be up to forty percent. It helps if you know that Doyle is only used for veneer logs due to the losses in milling. It also helps to know that Scribner is the scale most common in the Lake States. But neither of these facts will tell you how many board feet will come out of those trees. That will depend on the logger, how the trees are cut, what kind of equipment the sawmill uses, and most importantly, the experience of the sawyer at the mill. This is why there are three common scales. No one scale serves all products or situations.

To make an analogy, most people who invest in the stock market have a professional assist them with their investments. Likewise, it's an intelligent landowner who hires a professional to keep track of their investments in the forest and the market forces behind them. My forest can grow three to five percent a year in volume. Wood markets can change fifty percent or more in either direction during the same time. Accurate information is a key to success in any field. Forestry is no exception.

Minnesota Microcosm

I WATCH THE SEMI bearing down the highway outside my office window. It is filled with aspen "sticks." Ten cords of aspen are neatly stacked in eight-foot lengths in the cargo bay, headed for the pulp mill down the road. The aspen provides me with my writing paper. It also provides for hundreds of jobs for the economy of my community. Some of my friends work at the paper mill, others as foresters for the public or private sector. The forest is creating wealth and livelihoods in my neighborhood.

Many of my urban friends are surprised to learn what happens to the trees in our local forests. The numbers are interesting. In 1995 approximately four-and-one-half million cords of wood were harvested from the forests of Minnesota. That's a string of fireplace logs four feet high, sixteen inches wide, stretching four-fifths of the way around the globe at the equator. That's a long stack of firewood, but it's only two percent of the wood used by the population of the United States each year.

I look around at my office and home. I think about the quantity of paper in my life. As Americans, we consume more paper per capita than any culture in the history of humankind, ever! That's a sobering thought.

One third of the Minnesota wood harvest becomes wood pulp for paper products. Another third of the harvested fiber in Minnesota is ground into chips. These chips, about the size of a half dollar, are the basic building block for oriented strand board (OSB), the structural panelboard rapidly replacing plywood in the housing and remodeling markets. Billions of square feet of OSB are produced and shipped each year from northern Minnesota mills to markets in the midwest and the east.

The remaining third of the trees harvested in Minnesota go primarily for lumber or fuel wood. Only fifteen percent of the harvest becomes a traditional "board," and the majority of these are made into pallets for the shipping industry. In the end, a very small percentage, perhaps five percent of the wood harvested, is of high enough quality to make flooring or furniture or cabinets or paneling.

This fact bothers me. I would prefer to be growing tables rather than Sunday newspaper inserts. I would prefer that the products coming off my land had a long future ahead of them and not five minutes at breakfast and a lifetime in a landfill.

October Blues

I WAS BORN IN OCTOBER, so October should be a time of celebration. The early days of October offer magic to the eyes in the North Woods. The foliage on the deciduous landscape is a carnival of colors. Soft ochre on the birch, brilliant cadmium on the sumac, burnt umber on the red oak, Naples yellow on the maples. This palette undulates against the changing sky. It spoils my eyes with its glory.

Like all delights of the senses, however, the pleasure is fleeting. If my mind is on mundane matters, I will miss this show and awaken to the applause. If I follow the paper and wait until the weekend when the foliage is peaking, I will be disappointed by raw weather and leaves less magnificent than my imagination. The orgasm will have passed me by.

What follows the intense climax of color is a rapid decline of daylight. Post-coital depression, the ex post facto of leaf season, carries with it the winds of October blues. There comes a morning when the northwest wind bites at the back of my neck; a morning when there is hard frost on my windshield. These are two of the harbingers of change.

A third, more subtle difference is in the quality of light. All summer and early autumn the landscape has been bathed in green. Green is a soothing spectral color. It wraps me in my daily life, an oversized bath towel of warmth. I understand my neighbors who love the expansive fairways they call lawns. They try to maximize the quantity and quality of this light.

Now the leaves scoot and swirl past my windows. The green is passing. The hard rays of the sun reach the forest floor and shadows abound. The soft canopy of verdant foliage is passing away. I am greeted by the stark sky of winter.

Once again I become vulnerable to the wind and the night. I watch the leaves fall from the mysterious canopy. Where do they come from, these small sailing vessels, each a world unto itself? Some twirl. Some arc. Some plummet. I am reminded of the snowflakes on their solitary journey. Sometimes a gust brings a fleet of them bursting down against my window pane, scattering golden petals on the sill. But more often, they glide by alone, monks on a meditation through the air.

I watch the canopy thin imperceptibly, hour by hour. A sadness passes through my bones. In the morning it will be different. Heartache again fills my lungs. Separation and withdrawal return to my doorstep. Once again I will be alone with the sky.

My emotions are raw, like the late October air. A tear rises. The nests of the forest are empty and sway in the wind. I turn away toward my wood stove. It will soon be time for me to make my own heat for protection and warmth. The sound of an ax echoes in my mind. The October blues have come for a visit.

The Benefits of Blows

Blown Over

FOLKS IN THE MIDWEST speak of tornadoes with a certain quiet fear. There is an inflection in the voice, a look in the eyes. I had heard friends do this during my college years and had smiled inwardly, thinking silently that tornadoes were pathetic compared to hurricanes. Now I know better.

I experienced my first tornado in my twenties. It was a hot summer day with the typical late afternoon thunderstorm in the north country. My spouse and I had been out collecting the golf-ball-size hailstones with cookie sheets over our heads and were selecting the best ones to place in the freezer. What fun!

Then I happened to look out at the western sky. There was an unmistakably strange shape in the air—a dark cloud in the shape of a wedge. We knew instinctively what it was, and we headed for the root cellar. As I descended the stairs below the kitchen, I recall that distant roar. Yes, it did sound like a freight train. Then, all the doors in the house slammed simultaneously.

That day we were lucky. At the last moment, the storm veered to the south; the path of destruction spread just a quarter mile from the family summer home. One man died in that storm, and the memory has stuck.

Since that afternoon, I have had a different respect for the midwestern tornado. It is unpredictable and arrives without warning. There is no time to batten down hatches as there is in a hurricane. Not that it would do any good. One moment it is quiet; the next, there is raging chaos all around. And so I watch carefully for signs: the heavy air, the green sky, the hail. I do not give much attention to weather forecasts. I have been in two tornadoes now, and neither one of them was predicted by the forecasters. So much for their fancy color radar.

June 27, 1991, was another hot humid summer day in the woods. The weather forecast called for thunderstorms late in the day at Esden. Nothing unusual in that. I didn't pay much attention as I pored over plans for a silvicultural research project I was hired to jump start. On which sites could we get a bulldozer to do site scarification? Where were slopes too steep? What did the acorn crop look like? What kind of split plot design would we use? My mind was filled with thoughts other than weather.

Late in the day, I took a break and went for a swim. The air was heavy, as it frequently is before a front moves in. No heavy clouds in the west, though, and the forecast mentioned only the possibility of thunderstorms. It did not mention severe thunderstorms, or a tornado watch.

Usually, if we're going to get weather, it arrives late in the afternoon from the west. I can see it building out over the beaver hutch and the aspens beyond. This time there were no signs of anything—just clear, hazy sky. The evening passed without any storms materializing.

It was time for dinner, a book, and rest. The radio played music. It was after midnight and all the meteorologists were asleep. The night owl playing the music sat in a darkened studio. What did he know about the whims of Mother Nature?

At night in a small cabin on a ridge, I am alone. Being alone is nothing new for me. I am often alone. But on this particular night, I can't fool myself. I feel alone. One and a half miles from the nearest neighbor, the night is dark and foreboding. Rilke says it best in his *Duino Elegies*:

> *And even if I cried out,*
> *Who among that angelic order would hear*
> *me . . .*
> *For beauty's nothing but the start of terror*
> *we can hardly bear,*
> *and we love it because of the serene scorn*
> *it could kill us with.*

Rilke's reality is not an easy lesson. Much of my life is spent denying it, avoiding aloneness, but a windstorm will remind me quickly.

In the woods, on top of a ridge, thunderstorms have a certain presence. The lightning is the first to light up the interior. I have sat on my porch and watched this energy build from the west, build from an occasional whisper of light into a full fireworks display. There are brilliant, unpredictable flashes of energy, then blackness. I count the beats. No thunder yet. Another strobe on the windows.

It is a remarkable show of power. One moment there is darkness. The wall of black surrounds me in a cloak of absence. Suddenly the white strobe pulsates across the trees; the momentary illumination of a dream, vivid, harsh, remembered. Then absence returns.

The lightning is stronger now, incessant. I go to the window and stare at the sky to the west. I'm worried. This is unusual lightning activity with still no thunder. Opening the door to the west, it is calm, another bad sign. Where is the thunder? Where is the wind? And most of all, where is the rain? It is the rain that brings relief and breaks the spell. When the rain comes, I know that the worst is past.

Nothing prepares me for the wind. The first sound is a strange, unfamiliar, distant rumble. Puzzled by this alien clue, I step outside to gaze at the strobing blackness. This is a mistake.

Lured out of shelter, the gathering roar descends from the black void to the west. It is fast. It is furious. It doesn't take prisoners. With a sudden fiendish brutality, it approaches. I dash for shelter, knowing I have heard this sound before. "Damn," I whisper to myself, "Why me . . . why here . . . why now?" There is no one to help me. I drop onto my bed in a darkened corner and await oblivion.

The full force arrives with a vengeance on the blackened forest. My eardrums are overwhelmed by the cacophony of destruction. The house shudders at the first blow and starts to tremble like a canvas sail. The quivering increases and my mind carries the building up in the air and out to sea. The fury, the chaos, the darkness. Hollywood is no match for this creature of real anger.

₭ • ⌒

As I lay awaiting the final blow, the expectant shattering, the splintering flight into the void, my mind becomes calm and clear. If this is the time the great spirits have chosen . . . the human heart is small . . . but there is calm within.

For the first time I am aware of the variations in the wind's symphony. A chorus of groans, with altos and baritones, preside over the chaotic blast. The cries are long and mournful, rising and falling with the gusts in the eaves. There is a structure to this madness. . . . It endures and fills the emptiness with its cries.

Suddenly, my home ceases to shake. Before my ears can adjust, the rain hits, hard. This is not the soothing sound of gentle penetration. This is a hard percussive drumbeat. It pounds out of the skies and waves of water burst forth, driving themselves horizontally into the windows. I look up, but there is nothing to see. All is blackness and the winds have plastered every pane of glass with leaves and twigs. The chorus of moans begins to subside. I stare at the ceiling from my bed, not knowing if I am alive or in a dream or in the great beyond. I move my arm to my face and pinch my nose. Surprisingly, it hurts. I take a deep breath and breathe out slowly. I am awake and alive.

₧ • ₞

When the winds have passed, I carefully rise from my bed and venture into the living room with a small flashlight. The sound of water has replaced the wind, and some of it is coming from within. Pouring through an unfinished area of my west wall is a stream of rainwater. I swear quietly at the sight and round up a couple of pails to minimize the damage. Then I look for structural damage.

Directing my light at the large picture windows on the south side reveals a strange sight. The panes are a collage of leaves and twigs of oaks, birches, and aspens. It is solid-pressed green, flattened to obscure the darkness beyond.

I need to know more. The chaos outside has subsided. The wind and rain now carry the pitch of a summer storm, not of the furies of a few minutes earlier. Boots, rain gear, hat, these are close at hand. The ritual of venturing into the night with a small lantern is an old one for me. I know it well. Living without a functioning bathroom makes this a common journey into the unknown.

On this particular night my trepidation is higher than normal. As I step onto the porch the view is disheartening but not despairing. Beyond the shelter of the porch, one of my neighboring oaks has toppled across the path. Instead of relief that the tree has missed my home, I am thinking of my truck beyond. It is nowhere to be seen in the tangle of ghostly green in front of my eyes.

I venture forth from the sheltered porch, and splintered branches block my progress. The remnant strobes of lightning lend a surreal quality to the torn landscape. I push on around to the west. Then, as if Mother Nature had choreographed the blows, I spot my white truck. It is calmly untouched in a sea of surrounding chaos. From a distance it appears unharmed, with only small branches fluttering over the flat bed in back. There is a large oak down on either side of the vehicle. Relieved at last, I retreat indoors, still convinced that this was only a bad storm. I strip off the wet gear and collapse into a deep slumber. It will be morning before I learn the truth.

The Aftermath

WITH SUNRISE, I awaken to a new light. It is a clear calm morning: I slip over to the kitchen, my eyes catching glimpses of the sun flashing through the mat of green leaves that now call my windows home. This is a peculiar half light, and it reminds me that the night before was not a dream. Something has changed. It will not take me long to discover this.

Still calm, oblivious to the reality that lies outside my door, I am grateful to be awake. My cabin appears to have sustained only a few water leaks in the havoc. I will need to put off my research work for the morning and take care of the couple of oaks outside on my drive. My chainsaw is sharp and I find my chaps. I am used to a few trees falling on my mile-long driveway. The teapot is boiling. I pour a cup and, for the first time, glance outside my small kitchen window.

This window is protected on three sides and is the only one which has not encountered a new breed of leaves for company. The sky is blue. The sun is shining. All is well. But . . . something about the view is different. I pause, teacup in hand. The view has changed.

ഔ • ൙

And then it hits! *Everything* has changed. I turn to the door to verify my disbelief. Outside, on the porch, I again cast a glance to the east. There, across the small lake, a forest has vanished. Overnight, it has disappeared. Instead of a closed canopy of oak, dark shadows over the water, there is an open view for a mile to the east. In its place, a few scattered poles, striped of limbs, silhouette an open sky. The forest is no more!

I am speechless, stunned, and my legs are weak. I sit down and sip my tea in a trance of slow realization. My mind turns over the new landscape beyond and slowly focuses on the ragged remains of the forest. There are whole pine trees floating in the water. There are toothpick stalks bent and scattered around me. And most striking of all, there is the open sky. What has transpired in the night?

I rise and piece together the sounds and the images. This was not another fierce summer storm. This was a full-furied—I hesitate to say the word—tornado? What does this mean? How many acres are flattened? What is left of the forest I have been tending for ten years? The cough of chainsaws in the distance greets my ears. It is an ominous sign.

Suddenly, fearful for the first time, I jump up and dart out the screen door, forgetting my tea. What has happened? I rush past the oaks in an attempt to start down the driveway. But this is impossible. Torn branches and limbs block my path. I clamber over the tops and get stuck. I retreat and try again, further to the east. The jagged branches tear into my shirt. I duck and climb and straddle. Finally I arrive at my truck a full thirty feet from my doorstep. It is still nestled in a moment of calm. Instead of being thankful, I press on toward the driveway. A wall of green and gnarled limbs greets me.

For twenty minutes I clamber over and under, around and through a sea of torn trees. And with that I arrive at a small clearing, only fifty yards from my cabin. I am sweating and breathing heavily. I stop and look north, where my driveway winds through mature red and white pine. Only now, the pine are gone. Splintered stumps, split off ten feet in the air spike the view. I swear softly in anger. My blood rises. Why has nature dealt the forest this blow?

Later that morning, I will stumble and cut and sweat my way around these fallen beings on my way to my neighbors for help. I will count over one hundred and twenty trees down across my driveway. When I return with my neighbors, it will take three of us all day with two snarling chainsaws just to open a narrow swath to enable my truck to reach the road. That evening, exhausted and sweaty and hungry, I will retreat from this splintered landscape in frustration and fear. It will be months before I realize the extent of the damage to the forest—and a lifetime before the scars will heal.

Beyond the Furies

AFTER THE STORM, my relationship with the forest changes. The old woodland trails are closed. The cool shadows and shelter are gone, replaced by swarming insects. I break off a branch and strike randomly into the jungle. A wrenching feeling fills my stomach and lungs. My eyes glisten at the uprooted white pines. With my stick I strike a blow to the fallen oak on my left, a blow to the fallen pine on my right. I planted spruce here; I pruned oaks here; I thinned pines here. It was the shelter I sought away from the distrought world, a haven from the darkness. And now the darkness surrounds me.

The anguish is fresh. It is a splintered chaotic place. Where there was a stately oak, now a root ball lies twisted in its side with a gaping twelve-foot hole in the earth. Where there was shadow and calm, now there is hard light and wind. Where the path was gentle, now seventy-foot stems block my way.

The old forest is gone, broken, disintegrated. Before there was a known world, there was protection from the canopy over my head. Now there is naked sky, piercing light, shredded limbs and a cold wind blowing from the north. The fallen, green-cloaked branches are already starting to wilt in the darkening sky.

℘ • ℜ

As I clamber through the destruction, my mind tries to make sense of the storm. I begin to comprehend the forest in a new light. This new forest is not the bucolic place from my childhood. It holds all the passion and anger and willfulness of Kali, the Hindu goddess of destruction. She can turn her back as easily as she can caress. She can deceive as easily as she can love. She is as mean as she is kind. I step back onto my driveway, my shirt wringing with sweat, my jeans plastered with burrs. There is a lesson here, a lesson for the future. I am a romantic no more.

Salvaging the Past

ONE MORNING the salvage crew arrives with their machines. In the early light they begin their slow march of clearing and removal. I stand in the roadway and build brush piles behind them. A small bird alights on a branch nearby and starts to sing a mournful song. I torch another brush pile of debris and stare at the brilliant flames as the bonfire builds in intensity. A gentle moan courses through my throat. A salty trickle reaches my cheeks. I feel the moisture on my face. The salvage has begun.

To salvage the past means to reenter it after the storm has passed. It means to mark new boundary lines, to run a "cruise" of salvageable timber volumes, to lay out skid trails for the salvage crew. It means to become accustomed to the sound of a chainsaw making a sawlog out of damaged oaks.

To salvage the forest I need to move quickly. In the summertime the bark beetles are hungry. After six months my pines will be worthless. There are other pathogens on the prowl as well. Heart rot fungi will attack my aspen. In a year it will be mush. Amidst the deerflies and the mosquitoes, I try to measure what I have lost and what can be saved.

Salvage harvesting is challenging at best and dangerous at the least. There are trees leaning against trees leaning against trees. There are unexpected limbs that may snap back. There are hidden snags. The work proceeds slowly.

Despite the hardships, the oak logs build on the landing. Scaling stick in hand, I approach the fallen warriors. My fingers touch the closely grained log ends and rub sawdust away from the fresh cuts. My nostrils fill with the aroma of oak resins and tannins. The measurements are tallied in my ledger. I think of the oak flooring and boards these logs will make. The wood is strong and beautiful and durable. I run my finger across the end grain and wipe the sweat from my eyes.

As the salvage proceeds, the sadness of loss rides over me in waves. I try to ignore it and concentrate on the work at hand. I hear the buzz of chain saws as the crew moves into the heart of the forest. My memory recalls how it was before the storm. I see the oaks erect and firm. I remember the soft caress of the snow on the pines. I dream of the fluttering aspen leaves as I look for the trail where I spent hours walking in a meditative trance. These and other memories rise to the surface as the log pile builds on the landing.

Moving On

THE FURIES HAVE PASSED and rain falls from the sky. I look up and notice a pileated woodpecker feeding on a broken trunk. Twenty thousand feet of red oak sawlogs rest nearby, awaiting their journey to a new life.

I breathe deeply, in and out. Breathing out brings relief. Breathing in brings fresh air. My mind speaks. It is time to move on. Time to let go of the old forest that is no more. No amount of yearning will bring its return. It is time to think about replanting and reforesting the landscape.

After the salvage harvest is complete and the landings are barren of logs, I turn my attention to the topsy-turvy shrubs and soils. There is much to do to prepare for a new forest.

I begin with my friend on his yellow bulldozer. He returns down the same trail he created ten years earlier. This time his task is to clear the rubble the storm has left behind. Five acres of twisted trees and stumps are piled and buried. When the soil is dark and crumbles in my fingers I smile and wave to him. We are moving into the future.

There is a host of opportunities in the rubble that remains. There is a decade of firewood for my hearth. There are sawlogs of pine and oak to be sold for cash. There are maples and ash and aspen sprouting in the thickets of raspberries. There are thistles for the finches and raspberries for the bears. The world of acceptance is a world of possibility.

The following spring, I bring in a planting crew to replant the site to white pine and red pine. I protect the new seedlings with tree mats against the weeds. I protect them with bud caps against the white tailed deer. I cross my fingers and wait.

The Winds of Change

MY FORMER PARTNER, T., has a porcupine living under her summer cabin. "Porcolla," she calls her. T. and I disagree about Porcolla. This porcupine is slowly stripping the bark from all seven of the large oak trees around her cabin. The trees are suffering badly. If the porcupine continues its foraging, some of the oaks, perhaps most of them, will die. As a forester, I am partial to the oaks. I argue that the porcupine must go. T. argues that she must stay—"She's been here longer than I have," she declares. I shrug. I don't know where she gets her evidence of this, but you can't argue with a woman who's made up her mind. Logic is not part of the process. So, the porcupine remains, hovering over her head, nibbling away, killing the trees she loves.

Last week T. was painting in her studio. Three deer bounded by her window. She was surprised. The snow was deep, it was cold, and exerting that much energy in the middle of winter seemed unusual. Nevertheless, she watched with wonder at the display. A few seconds later, five more deer followed- with the same urgent run and leap. There were two small fawns in the center of the group. It was a spectacle. T. watched out her window, puzzled.

Thirty seconds later, the answer appeared. A large canine, gray and bushy tailed, ran along the trail the deer had made. T. was upset by this. She went outside. "Bad dog . . . go home," she called out into the silence.

She recounted the story to me on the phone. I paused and smiled. "Are you sure this was a dog?" I ask. She grew silent. I could feel her eyes widen. "A wolf?" she whispered back across the line. It was a wolf, a wolf looking for a meal in winter.

ಬಿ • ಲ

Is there a lesson here? I sometimes wonder how long Mr. Thoreau spent in the woods. He writes with eloquence and grace, but the landscape he describes is only a portion of what I have experienced around me.

Perhaps in his short tenure fifteen miles west of Boston, he observed only the rejuvenating songs of the forest in the springtime. Perhaps he wrote down only selected observations to relieve himself of the burdens of urban life. Perhaps he did not see the dead and dying trees around him.

It is clear, as I sit staring at the holocaust the windstorm left behind, that Mother Nature has a cruel streak. I may sit down to rest on a hot summer day and be stung into flight by the wasp nest at my side. I may be entranced by the trillium in full bloom and ignore the deer tick under my sock. In the morning, I will awaken to swelling and itching and infection. In the winter, I may linger too long in a brisk wind and frostbite will find my exposed cheeks.

Certainly, there are days when the scarlet tanager sings brightly from the dogwood or when the soft-orange sugar maple leaves drift gracefully to the ground. There are days when the smell of the balsam needles between my fingers triggers fond memories of a childhood Christmas.

There are also days when the white pine seedlings I planted turn yellow and brown from blister rust infection, days when the industrious beavers girdle my favorite red oak along the pond walk, days when the forest droops from the munching of ten million caterpillars defoliating her greenery. These are not isolated events. They occur with regularity and they lend credence to the theory of disturbance ecology. The stable "climax forest" is a rarity. Sometimes it is humans who make the disturbance with our chain saws. Perhaps more frequently than most of us realize, it is the furies of nature—wind, disease, or fire—which upset, disrupt and dismember the forested landscape.

ഇ • ര

The forest, like the heart, heals slowly. There is no instant medicine that alleviates the pain, no quick remedy that relieves the heartache. The forest is emotionally and physically exhausted. It longs for a cool breeze, warm sunlight, soft evenings and peaceful rest. The forest, like my heart, wants to go to sleep. It wants to rest its weary eyes and let time slowly pass over its face and dry its tears. Finally, there is peace.

Sustainability and the Future

Beginnings and Endings

WHAT IS THIS WORLD OF NATURE I call home? The tornado opened my eyes. I survey the damage five years later as life again takes hold. Out on Esden Lake, there is a line where the storm tore through. I paddle out and let the wind carry me. To the north a mature oak forest is intact. It is tall, stately and quiet. Barely a whisper emerges from its darkened understory. My canoe drifts south toward the storm line.

Only a couple of hundred yards from the oaks, twisted trunks still dot the shoreline. Barren branches rest in the water, slowly rotting. Dead snags silhouette the inland landscape. Tortured aspen are bent like linguine on a spoon. Beneath this tangled madness, a chorus of songbirds sings in the morning light. The brush is alive with a cacophony of sound. It is here, in the protected brush and shrubs, that life takes hold anew. Meanwhile, the oak forest to the north is silent.

A barred owl alights on a snag in the midst of the storm damaged aspen. I pause and stare at her. She does not take flight. She twists her neck and stares back. My canoe holds its breath. We sit this way, cautiously eyeing each other, for at least five minutes. These are shy birds I do not often see close-up. It is a treat for me. I'm not sure whether she feels the same way. Do we understand each other?

I am pleased that the owl has returned to the ravaged woods after only five years. Is she pleased that I have returned? Can we live together on this place? My mind is full of questions. Questions about the paradoxes around me. The mixture of life and death in the forest puzzles. I recall the scene in *Anna Karenina* where Levin's brother is dying his agonizing death from consumption. His wife, Kitty, becomes ill during the last watch. Death seems to be everywhere. And yet, as we discover later, her illness is the first sign that she is with child. Tolstoy understood what I am only beginning to accept.

Actually, the owl and I are similar in many ways. We both are observers, sitting in our perches watching and waiting. We are also both solitary sorts. I won't find the owl in a colony. She lives alone in the woods. That is my preference, too. And finally, the owl and I are both predators. What did I have for lunch? She lurches out a fur ball of woodland residue, and silently takes flight into the oaks. She is dependent on this forest. I am, too. If she leaves, I will not be far behind.

Competition for the Light

IN THE FOREST there is only so much room in the ground. Every herb, every forb, every shrub, every tree lands on a site as a seed or seedling or root sucker. It begins to draw energy for its own survival. Aboveground, the competition is for light. Below ground, the competition is for moisture and nutrients. In both worlds, the competition is intense, particularly after a disturbance.

Human beings adapt with a variety of strategies to the environments we colonize. Trees are no different. Sugar maple is an example. Adapted for the shade, it seeds on cool, moist soils of fertile sites. Quietly, it drops its seeds in the shadows beneath the ferns and the oaks. Acorns fall next to it, germinate, grow into small seedlings, and then wither from the lack of light. Likewise, the birch seed, the pine cones, the aspen suckers, all perish in its shadows. But the maple will hover near the ground and slowly cover an entire site with its offspring. If we let it, it will take over a site.

Competition takes many forms and sugar maple is only one species in the forest competing for space. Each species has its own adaptive strategies for survival. The most aggressive species, like jack pine, birch, and aspen, colonize a site quickly and shoot for the sun. These are the ones that arrive early, land on mineral soil, and grow more rapidly than their other hardwood cousins. They are the pioneers and keep their heads above the competition for the light. If I happen upon a pure birch stand on a stroll through the woods, I am witnessing a pioneer stand. Birch settled the site following a severe disturbance. What was the disturbance? Perhaps it was fire. Perhaps it was farming or a windstorm. There is continual struggle occurring in the forest as there is in human culture.

After the windstorm, my forest starts over. In places where the light is strong, the aspen rises. Where some of the oaks survived the winds, their bent boles shelter new maple seedlings. And where my hand cleared the debris, small pines reach for the sky.

A Sustainable View

I AM SEATED IN A CHAIR at my home on Esden Lake. It occupies the same place on the map that it did fifteen years ago when I put in my road, yet it has changed completely. In a forest, there is no such thing as stasis, except perhaps on a January morning when the thermometer reads forty below zero and all is still.

There are many today who believe that the forests of old can be preserved. They cannot. Preserving a forest is tantamount to embalming her. Perhaps we wish to reserve some older trees, so that they might have a longer life span. This we can do, and probably should. But we cannot preserve a forest. The sooner we understand this, the healthier all of us, including the forest, will be.

When we choose to reserve older trees in a forest, we encourage different changes to occur than when we harvest them. Certain species—those which are shade tolerant, like sugar maple and hemlock—will find their progeny increase dramatically. Others—those that are shade intolerant, like Douglas fir, red pine, and trembling aspen—will find it difficult to regenerate their future generations. We will have reserved older trees at the cost of limiting their offspring. This is not preservation; it is change.

We might remember that foresters were this country's first conservationists. Long before it was fashionable, foresters went to the woods to stop the destructive fires, limit the exploitation of western forests, and replant thousands of acres of cut-over and burned eastern woodlands. It was foresters, like Gifford Pinchot and John Muir, who saw the bigger picture and pushed for changes in management and philosophy.

Today we need more foresters in the woods, foresters who are trained to communicate as well as to count "MBF." We need foresters who understand ecological habitat typing as well as aerial photographs. We need foresters who see the big picture as well as the small picture.

Harvesting in our forests is no more a misguided practice than is harvesting in our gardens. There is a great deal of discussion today about sustainability of resource systems. The dialogue is a first step toward addressing a critical concern: depletion of our natural resources. Almost everyone agrees that it is a foolish and a dangerous thing to deplete our natural resource systems, and yet we are doing precious little to change our lifestyles.

When we decided to forbid most harvesting in the western public forests due to legitimate concerns about wildlife habitat, did we decrease our consumption commensurately? Not at all. In fact, in the past ten years our consumption of wood and paper has increased. As a result, the United States, which used to be a net exporter of forest products, now imports more fiber from Canada, Chile, New Zealand, even Russia. So much for our balance of payments. We want to have our cake and eat it too.

What does this mean? It means that, while we are addressing the sustainability of our national forests by reducing the harvest of mature timber, we are increasing the strain of harvest on world forests. Are we aware, for example, of the increasing harvest levels in the Russian Taiga, a large wilderness area, which produces tremendous quantities of oxygen for the planet?

It is commendable for us to begin to discuss sustainability. But discussion is only a first step. If we do not act responsibly, sooner or later Mother Nature will exact her revenge, and the price will be great. Where do we begin?

೫೦   •   ೧೪

It turns out, there is a simple way to measure forest sustainability: remove from the forest no more than it is currently growing. For example, the northern hardwood forest in my county averages one-third of a cord per acre per year of growth. If we harvest less than this level on a county-wide basis, we are practicing sustainable forestry. The United States Forest Service has excellent records of the growth rates for forests in all parts of the country. This is a good place to begin. Over time, as we learn more about the capacity of our forests, we may be able to adjust the harvest level. But at first, we would do well to err on the conservative side.

There are other methods of measuring sustainability on a local level, but we need accurate information for these to be effective. For example, if I know from recent inventory that I have a thousand cords of oak in my forest and I also am reasonably certain that I can grow it again on a one-hundred-year rotation (one crop every hundred years), then it is sustainable to harvest an average of ten cords a year from this forest (one percent of the volume per year).

These are not calculations that require algorithms, calculus, or computer programs—although those can be used to refine them. Instead, they are simple calculations all of us can make, simple calculations for sustainable forestry. They provide a place to begin to think sustainably. This way we may enjoy our hardwood floors and our Sunday papers and know that we are leaving a healthy future for our grandchildren.

The Power of Silence

IT IS FEBRUARY SECOND and cold today in Minnesota. Very cold. The thermometer this morning was off the deep end. Literally. The readings stop at minus forty, and it was below that. It was colder here than at the North Pole. And, in Tower, Minnesota, they had a new state record: minus sixty degrees Fahrenheit.

Nevertheless, I joke about it to stay warm. I tell my friends in southern California that they should visit. I remind others close to home that the benefit of thirty below zero is that it will feel warm when we see zero.

Actually, in Minnesota we know the difference between twenty and thirty below zero. This morning, February 4, it is warmer—only minus twenty-two. Yup, a real warm front.

The winter weather here makes me do the same thing that southerners do in summer: stay indoors. They insulate their houses to stay cool. I insulate to stay warm. In winter they are outdoors a lot. In summer I am outdoors a lot. It's pretty much the same thing if you ask me.

What is most impressive about the bitter cold is the silence it carries on its back. Step outside at thirty below and listen to the silence—no automobiles, no snowmobiles, no outboard motors . . . silence. Do you remember what it sounds like? I sometimes forget.

Silence is powerful. The snow insulates and muffles the sound of humanity. All is still. Last night a new blanket of white powder fluffed down on top of the couple of feet already on the ground. As long as I don't have to drive, it's serene and beautiful.

The crystalline sun spreads its rays across the winter landscape. Puffs of snow are swept from the balsams by the wind. The low, long light offers shadows and colors as it circles the birch, the maple, the aspen.

I slip on a pair of snowshoes and head out into the whitened world. My parka is curled around my face. Only a small sliver of light reaches my eyes in the biting wind.

The snow is deep. Even with snowshoes, it is work to blaze a trail into the pines. It will be easier on the return. As I walk, my legs find a rhythm, steady and slow. Puffs of snow sprinkle over my head as the wind blows the powder from the evergreen bows. I turn away from the north wind and watch the snow shower. The wind is singing a soft song of bitter cold. It's a little like the echo of the ocean waves from a distance, a low roar with pulses.

Mine are the first tracks in this part of the forest. The smarter creatures are all hiding out beneath the snow, using the fluffy down as insulation from this cold and wind. But I was getting cabin fever. I need the exercise.

Exercise is one of the advantages of snowshoeing. It's a little like cross-country skiing, but without the arm motion and the rigid trails. Here, I make my own trail. My heart beats faster as I spot a white pine I've never seen before. Funny, I've walked this way many times, and still the obvious is not always apparent. Where did that white pine come from? Has it always been there? It's mocking me for my blindness as I pass.

Later, I return through the pines to the trail. As I do, a dull engine stirs the silence. Two, then three snowmobiles appear. They are carrying dark, hooded travelers. I stop as the phantoms growl past. They lift their right arms in a salute to the solitary walker.

As they pass, my ears ring, and my nostrils fill with the smell of burning hydrocarbons. I watch them disappear around the corner, and the silence returns. Funny, that tradition, the sitting, the noise, the smell. I've never been on a snowmobile and have been thinking that I should try a ride sometime, just to see what the appeal is. Perhaps it is the speed; in our culture speed is everything.

But then I think of the drawbacks. I can't leave the trail in deep snow, or I'll stall. My exercise will have to wait. My nose will fill with hydrocarbons. And most of all, my ears will be closed to the silence. It is this last factor that causes me to shake my head and postpone the urge to hop on. I keep on walking. The slow *swoosh* of white ash under my feet reassures me that I have made the right choice.

I respect the silence, this simplest of senses. The calm it brings. The peace of mind it offers. It is one of the reasons I enjoy fieldwork. Now, in the depth of winter, I can compare the real thing. It almost hurts the ears it is so powerful.

A *rat-a-tat-tat* interrupts my mindful musings. I glance across in the direction of the intrusion. There is a dead aspen snag fifty yards away. It is suspect. A bright red head appears around its side. *Tat . . . a . . . tat-tat. Tat-tat.* The large shape moves in the winter light. The pileated woodpecker finds no fear in this winter cold.

Waiting for the Sun

I AM WAITING FOR SPRING, waiting for the sun to rise in the eastern sky and cast a warm glow to the landscape. Waiting for the white, cake-laden boughs to cast off their icing and show their evergreen beneath. Waiting for my mailbox to appear again from underneath its three-foot white blanket.

Foresters are trained to wait. I plant trees that I will never see full grown. I thin forests that the next generation will harvest. Still, sometimes in February, the waiting is interminable in the north country.

There are hints of warmer times on the horizon. The temperature rises above thirty-two degrees for the first time in two months. The roof comes alive with the sound of dripping water. Liquid pools form in the driveway. The sap starts to flow.

But it is only a tease. Three days later, under a new addendum of white insulation, the northwest wind blows over with wind chills in the danger zone. I look out over the frozen landscape and wonder if the sounds I heard a few days earlier were only in my dreams.

And so I wait. Although I am trained to wait, it is not always easy, especially this time of year. February 27th. The rest of the country seems to be fairly basking in warmth—at least according to the wonderful color-coded isobar weather maps that flash across the local television screen. Forties, fifties, sixties: are they making this stuff up? Does the thermometer really need to have any readings above thirty degrees? For whom? All those lucky folks south of the Mason Dixon line, I suppose.

Just what does fifty degrees feel like? I've forgotten. Long underwear is my second layer of skin. And there are signs that I'll even have to start rubbing moisturizing cream on this chapped pair of gloves at my side. Ah, the pleasures of late winter. I sit back with a hot cup of tea, gaze at the frozen world outside the window, and wait for spring. It will not be long.

ℬ • ℭ

Henry David Thoreau was not all wet. There is romance in the forest. There is a time in the early spring when the woodlands send forth the powerful aroma of youth at its physical and sexual peak. The air stirs with electricity.

There is a time when a warm breeze blows in from the south; a time when sunlight reaches the forest floor; a time when the air is caressed with the gentle laughter of a migrant scarlet tanager. He arrives with a flourish, garbed in stunning colors. It is the season of the wild plum and the *Amalenchier* blooming. The fragrant, fragile flowers are harbingers of new life on old limbs.

The physique of spring is exquisite, well-toned and taut with power. Amidst the fanfare, the forest responds. The dormant cells swell and release the green buds into the sunlight.

This is a time of magic and mystery. The birds and the bees perform their duty of rejuvenation and the sap rises sweetly. The hepatica on the forest floor, the ferns, the trillium—these are among the first to feel the heat and burst into color. This is their moment of sunlight, before the green canopy closes over their heads. Dried leaves are pushed aside. A sweet perfume fills the forest. It is the perfume of rebirth.

An hour, a morning, a day spent traversing the woodlands during this window of romance is a blissful time. To be in spring's presence, this alone is enough. The imagination makes the leap, and the heart fills. There is lightness in the air, exuberance in the spirit, a smile in the eyes.

Early spring carries a clarity of purpose on her shoulders. She is focused and sure. She knows where she is going, and she carries the future with her. I am blessed in her presence.

The Quality of Light

WE HAVE TRAVELED FULL CIRCLE on the road through Esden Lake. We have been introduced to the forest and its paradoxical ways. We have planted and harvested. We have experienced growth and destruction. We are preparing to live sustainably as we are exposed to the light of the woods.

Sometimes the light at Esden Lake is soft and full of mist, with the vapors rising off the water in slow motion. The undulating curves drift over a primordial time while the resident loons call mournfully in the background. It is a peaceful, surreal light. There are no shadows, only ghostly figures and silhouettes rising into a timeless space.

Sometimes the light at Esden Lake is sparkly and bright. Ten thousand points glitter off a surface of blue. The sun is brilliant and sharp, with smoke-puff clouds gliding past. It is a cheerful time. The chipmunks chatter, the yellow warblers flit furtively in the hazel and birch. The oaks are radiant in their green cloaks.

Sometimes the light is moody and dark and gray. The low rain clouds sulk in from the west, blowing the moisture before them. The light is somber, muted in the sadness of umber and charcoal. The woods are silent. The leaves drip. The wind is bitter.

And sometimes, the light is dramatic and wild: a sunset pattern of crimson and violet and gold in the western sky; a cold front gusting hard with frayed leaves from the north, an evening moon growing full with reflection over the lake to the east; a wall of thunderclouds ominously darkening from the south.

There are moments of drama in the skies around me and moments of calm. I watch them with pleasure and with fear, for they are larger than I am and have changed my life. Like a sponge, my bones have soaked in the moods and the magic. What can I offer in return?

I am on the road to a new relationship with the forest. It is a relationship based on understanding and acceptance. If I experience the forest as a sustaining relationship, then the two of us are partners. She is not superior. I do not place her on Thoreau's pedestal. I am not superior either. I do not control her. We work together, live together, grow together and die together. I respect her secrets and she respects mine. I listen to her and she listens to me. She knows I need wood to keep warm, space to breathe, paper to write these words. I know that she needs a fertile seedbed, moisture and sun, and room to grow.

There will be disturbances in our new relationship. Sometimes she will have her way. Sometimes I will have mine. But we will return and help each other.

This is a new journey. For it to succeed we must both be sustained. It will not be an easy trek.

There will be obstacles in the path. There will be those who try to discourage or destroy us.

I will have to leave old relationships behind. I will feel alone, abandoned, heartbroken. But the journey will be worth the losses. The air will invigorate me. Beyond control and consumption, beyond romanticism and preservation, we will grow toward understanding and respect. But first we must let go of the past . . .

The End

Acknowledgements

Many individuals contributed to help bring this book into being.

First of all, I, like most writers, am deeply indebted to my fine editor, Corinne Dwyer, at North Star Press. Corinne has been patient, persevering and persuasive as she guided the manuscript and the author through a steep learning curve.

Secondly, the book would be missing a key link without the wonderful, dramatic illustrations by Mary Sandberg. Mary's eye, her sensitivity to the text, and her finely honed graphic skills give the pages added depth and charm.

Many others also gave their time and thoughts to the manuscript at various stages of development. Early readers who offered encouragement include, Dave Skeie, Jack Goldfeather, Marion Abbott, and Frannie Howe. Scott Edelstein provided excellent literary advice to help me expand many of the shorter essays. Rod Jacobs and Greg Nolan, both fine forestry professionals, helped me keep the focus clear and the facts straight. Finally, Edith Wright and Ali Kahn were sharp copy editors in the late stages, seeking carefully crafted language and a polish for the final proofs.

To everyone who put up with me during the process, especially my wonderful and supportive son Shawn, thanks again.

Glossary

Auger:
A power tool used to drill cylindrical holes in the ground prior to hand planting.

Basal area:
The cross-sectional area of a tree at breast height.

Biodiversity:
The biological diversity of a given area, whatever its size, is a measurement of the variability of species (flora and fauna) found on that site.

Bolt:
An intermediate-sized, 100-inch-long piece of wood, usually eight to twelve inches in diameter.

Cambium:
The live layer of tissue in trees between the inner bark (phloem) and the wood (xylem).

Canopy:
The upper strata of a forest.

Climax forest:
In classical ecological theory, a climax forest is one that will reproduce itself ad infinitum in the absent of any out-side disturbance.

Cord:
A measure of wood volume equal to seventy-nine cubic feet of wood.

Crop tree:
A forest tree that is selected to grow to maturity due to its vigor, form, and position in the canopy.

Crown: The three-dimensional, live growing surface of the top of the tree. Usually trees are classified by their crown position as supressed, intermediate, co-dominant or dominant.

Cruise: The act of conducting an inventory of tree stands.

Cull tree: A poorly formed or weakened tree marked for removal.

Current annual increment: The growth of a tree for a specific year, usually expressed by the width of tree rings at breast height.

Disturbance ecology: The theory of ecological response to landscape level disturbances—how forested ecosystems may react to windstorms, fire, insect outbreaks, disease, or harvesting.

Driftless area: The portion of southeastern Minnesota, southwestern Wisconsin, and northeastern Iowa that escaped the last glacial ice advance and hence does not contain glacial drift.

Ecology: The branch of biology dealing with the relations between organisms and their environment.

155

Habitat clas-
sification: A method of classifying and mapping forested areas by the associations of plant communities found there.

Hoedad: A flat-bladed planting instrument with a long, pick-like handle.

MBF: One thousand board feet of wood. One board foot is a piece of wood one inch thick, twelve inches long, and twelve inches wide.

Natural
regeneration: Reforestation of an area by other than human hands.

Plantation: A forested area where the majority of trees were planted by humans.

Primary
succession: The natural replacement of an unforested area by a forested cover type (e.g. prairie to woodland).

Scaling stick: The instrument used by a forester to measure wood volumes.

Secondary
succession: The natural replacement of one forested type by another forested type over time.

Shelterwood: A management technique for establishing a new forest stand under the shelter of the partial canopy of the prior stand.

Silviculture: The art and science of tending a forest.

Silvicultural
prescription: The method of attaining a desired condition for a forested area.

Site preparation: Silvicultural prescriptions used to prepare a site for planting. Examples include furrow plowing, mechanical soil scarification, or application of herbicides.

Stick: A one-hundred-inch-long piece of wood. This measure usually is utilized for pulpwood and wood between five and eleven inches in diameter.

Stumpage: The value of standing trees in a forest.